MEXICAN SCHOOL OF PARTICLES AND FIELDS

AIP CONFERENCE PROCEEDINGS 143

RITA G. LERNER
SERIES EDITOR

MEXICAN SCHOOL OF PARTICLES AND FIELDS

OAXTEPEC, MÉXICO 1984

EDITORS:

J. L. LUCIO & A. ZEPEDA
CENTRO DE INVESTIGACIÓN Y DE ESTUDIOS AVANZADOS DEL IPN

M. MORENO
INSTITUTO DE FÍSICA, UNIVERSIDAD NACIONAL AUTÓNOMA DE MÉXICO

AMERICAN INSTITUTE OF PHYSICS NEW YORK 1986

L.C. Catalog Card No. 86-81187
ISBN 0-88318-342-0
DOE CONF-8412105

Printed in the United States of America

Table of Contents

Selected Topics in Gauge Theories 1
M. A. B. Bég

Hadron Colliders, Intermediate Vector Bosons, New Quarks, and Leptons 58
D. Cline

Supersymmetry, Supergravity, and Particle Physics 97
B. Ovrut

Composite Models of "Elementary" Particles and Fields 122
H. Terazawa

Neutrino Masses 171
D. Wyler

List of Participants 261

List of Seminars 263

FOREWORD

The First Mexican School of Particles and Fields was held in Oaxtepec, Morelos from December 3 to December 14, 1984 and was devoted to topics in the frontier of the theoretical and experimental physics of elementary particles.

It is a pleasure to thank Professors M. A. B. Bég, D. Cline, W. Marciano, B. Ovrut, H. Terazawa and D. Wyler for their excellent lectures. Their lecture notes, with the exception of those of Professor Marciano on the Standard Model, form the contents of these proceedings. In addition, we organized discussion sessions and accepted contributed talks of participants.

The school was made possible by the generous support of the following Mexican institutions: Consejo Nacional de Ciencia y Tecnología, Secretaría de Educación Pública, Universidad Nacional Autónoma de México, and the Centro de Investigación y de Estudios Avanzados del Instituto Politécnico Nacional. The U.S. National Science Foundation provided the funds for the publication of these proceedings. In addition we would like to take this opportunity to express our appreciation and thanks to all those who helped to make this school possible. We would especially like to acknowledge the work of our school secretaries, Ma. Esther Vargas, Pilar Cassy, and María Eugenia López.

J. L. Lucio
A. Zepeda
M. Moreno

Oaxtepec, México
December 1985

ORGANIZED BY:

Sociedad Mexicana de Física

SPONSORED AND SUPPORTED BY:

Consejo Nacional de Ciencia y Tecnología

Secretaría de Educación Pública

Instituto de Física
Universidad Nacional Autónoma de México

Departamento de Física
Centro de Investigación y de Estudios Avanzados del IPN

U.S. National Science Foundation

ORGANIZING COMMITTEE

R. Huerta

J. Keller

J. L. Lucio

M. Moreno

M. A. Pérez

L. F. Urrutia

A. Zepeda

SELECTED TOPICS IN GAUGE THEORIES

Lectures Delivered at the Escuela Mexicana de Partículas y Campos, Oaxtepec, Morelos, México. December 3 - 14, 1984

by

M.A.B. Bég*
The Rockefeller University, New York, N.Y. 10021

Written version prepared in collaboration with

A. Hernández Galeana**
Centro de Investigación y de Estudios Avanzados del IPN, 07000 México D.F.

*Work supported in part by the Department of Energy, under Contract Number DE-AC02-81ER40033B.000, and the National Science Foundation.

**Also FESC-UNAM.

CONTENTS

1. INTRODUCTION

2. CANONICAL DESCRIPTION OF ELECTROWEAK AND STRONG INTERACTIONS
   - 2.1 Basic Principles of Gauge Theories
   - 2.2 Flavor and Color Spaces
   - 2.3 The $SU(3)_C \otimes SU(2)_L \otimes U(1)_Y$ Based Standard Model
   - 2.4 Symmetry of the Higgs Sector
   - 2.5 Status of QCD
   - 2.6 Status of Standard QFD
   - 2.7 Interface of QFD and QCD

3. HIGGS MASS IN SALAM-WEINBERG THEORY
   - 3.1 Some Rigorous Results
   - 3.2 Heuristic Considerations
   - 3.3 A Framework for Resolution
   - 3.4 Renormalization Group Analysis and Results
   - 3.5 Extensions of the Scenario
   - 3.6 Questions of Principle
   - 3.7 Concluding Remark

4. MORE ABOUT THE SHORTCOMINGS OF THE CANONICAL FORMULATION, QUESTIONS OF ELEGANCE AND TASTE

5. DYNAMICAL HIGGS MECHANISM
   - 5.1 Introduction
   - 5.2 Flavor Symmetries in QCD

6. HYPERCOLOR THEORIES
   - 6.1 The Hypercolor Scenario: An Artist's Conception
   - 6.2 Problems and Tentative Solutions
   - 6.3 Quark Current Masses
   - 6.4 Flavor Changing Neutral Currents
   - 6.5 The $\Delta I_{wk} = 1/2$ Rule
   - 6.6 Phenomenological Implications of Hypercolor
   - 6.7 Present Status of Hypercolor

7. COMPOSITE MODELS OF QUARKS AND LEPTONS
   - 7.1 Motivation
   - 7.2 The Quark Model Paradigm
   - 7.3 Composite Leptons and Quarks
   - 7.4 Hyperfermions as part of a Subquark Multiplet

8. OUTLOOK

9. REFERENCES

1. INTRODUCTION

During the last fourteen years, the conceptual framework underlying our thinking about elementary particle phenomena has undergone a substantial change. That Gauge Field Theory[1], realized in (1+3)-dimensional space, provides the most economical and elegant description of all interactions -- other than gravitation, for reasons (and in the sense) specified below -- is now universally recognized. The crucial factors that led to this "gauge theoretic revolution" were the theoretical requirements of renormalizability for the weak interaction[2] and of asymptotic freedom for the strong interaction[3].

It was realized, early, that the problems of describing the weak interactions in a renormalizable way and of synthesizing them with the electromagnetic interactions could be tackled together; indeed the principal bonus from this synthesis, the realization that weak interactions owe their weakness to the heavy mass of the mediating fields, continues to have a profound impact on the thinking of many physicists at this time. Since the known structure of weak interactions at low energies mandates a weak gauge group containing $SU(2)_L$, the solution chosen by Salam and Weinberg for the unifying electroweak group[2], to wit $SU(2)_L \otimes U(1)$, is the simplest consistent with observation. A history of the discovery of this group is a fascinating subject in its own right; suffice it to mention here that the first uncertain steps, in what we now know to be the proper direction, were taken in the late '30s by Oscar Klein[4] -- in a prescient paper which contained the germs of the Yang-Mills theory[5] and anticipated many other recent developments.

To justify a specific gauge group for describing the strong interactions, it is necessary to go back to the days of SU(6). We need only recall the order put by this group into the low-lying baryon spectrum[6,7] and of its implications for baryonic magnetic moments[8]. All of these results stem from the assumption that the wave-function of the three quarks inside the baryon is completely symmetric under permutation of the known degrees of freedom; consistency with the spin-statistics theorem then requires a new trichotomic degree of freedom,[9] dubbed "color" by a number of authors. Quarks of any given flavor are thereby tripled, and the underlying Lagrangian acquires an additional global invariance under a group SU(3). Gauging of this degree of freedom emerges as a tempting way to introduce strong interactions in a manner which is asymptotically free; the theory known as Quantum Chromodynamics[10] (QCD) is thus conceived. The principal virtue of QCD is that -- apart from CP-invariance, an exception that leads to the strong CP-problem[1] -- it embodies all the known symmetries of the strong interactions in a way that is both automatic and natural. Thus the strong, electromagnetic and weak interactions of all known elementary particles, at least in the low energy domain, may by derived from a gauge principle predicated on the group:

$$G_{L.E.} = SU(3)_C \otimes SU(2)_L \otimes U(1) \qquad (1.1)$$

To give masses to the fields mediating the weak interactions, the $W^{\pm}$ and the Z°, in a manner such that the twin requirements of renormalizability and unitarity are not jeopardized, it is customary to invoke the Higgs mechanism[11,12] and trigger it with the help of scalar fields tranforming as a complex doublet under $SU(2)_L$. This corresponds to four real fields; with properly chosen parameters one obtains three massless fields, Goldstone modes which combine with the appropriate gauge fields to yield massive $W^{\pm}$ and Z°; one massive field is left-over, to bear witness to this realization of the "symmetry breaking" process. The model of elementary particles based on the group $G_{L.E.}$, with this mechanism built-in, and characterized by three-quark and three lepton families (making a total of 24 fermions, as indicated in Table 1.1) has come to be called "the standard model".

ν_e	ν_μ	ν_τ	
e	μ	τ	
u_i	c_i	t_i	
d_i	s_i	b_i	$i = 1, 2, 3$

Table 1.1 The presently known elementary fermions. ν_ℓ, $\ell(\ell = e, \mu$ or $\tau)$, the neutrinos and the negatively charged leptons, carry unit lepton numbers; u,c,t (d,s,b) are the charge +2/3 (-1/3) quarks, carrying baryon number 1/3.

Within the framework of the canonical methodology -- that is to say, the use of elementary scalars to initiate the Higgs process -- it was soon realized that if a physical principle could be clearly spelled out, one could almost always also incorporate it in a gauge theory -- with judicious choice of the gauge group and the Higgs sector. This simple truth underlay many of the subsequent developments, and the resultant plethora of publications.

To appreciate the significance of the next step in the unification program, it is good to recall that Fitch and Cronin[13] decisively established CP-nonconservation in 1964; the most interesting implication of this phenomenon was deduced by Andrei Sakharov[14]: he noted that if b, the baryon number, is not a constant of the motion, then CP and b nonconservations could act in concert and produce the observed matter - antimatter asymmetry -- within the framework of the hot big bang theory -- by starting from a symmetric initial state in which matter and antimatter exist in equal amounts. This observation lent such elegance to the standard model of cosmology (not to be confused with the above-described standard model of elementary particles) that the search for a gauge theory which could (a) accommodate CP-noninvariance and (b) yield baryon and lepton nonconservation became a worthwhile quest. To accommodate CP-violation is neither difficult nor, at least at the moment of this writing, terribly enlightening. However, the momentum generated by the electroweak synthesis played a crucial role in meeting the second part of "Sakharov's challenge"; it led to several proposals for a further synthesis, the embedding of electroweak theory (often called Quantum Flavordynamics or QFD) and QCD in a gauge theory based on some suitable Lie group of rank ≥ 4. The simplest solution to this embedding problem[15] is afforded by the group SU(5), and it leads to proton decay.

The grand-unification group SU(5) is presumed to become an exact symmetry at energies of the order of 10^{14} GeV, where the three running coupling constants in $G_{L.E.}$ become equal; at this "triple point" the Weinberg angle becomes a purely group theoretic parameter, and may be shown to be such that $\sin^2\Theta_w = 3/8$; extrapolating down to low energies[16], we find that the observed $\sin^2\Theta_w \simeq 0.22$. Grand unification thus adds another successful prediction to the many triumphs of the electroweak synthesis. From a purely theoretical viewpoint, the embedding of $U(1)_Q$ the electromagnetic gauge group, in a simple group such as SU(5), renders Quantum Electrodynamics logically consistent in an unambiguous way. About this, we shall have much more to say later.

A very similar story can be told about a subject of lesser import, the recurring conjecture that the universe is ambidextrous, even though weak interactions appear to prefer left to right. The first clear formulation of this principle is due to Lipmanov[17], who pointed out that a left-right symmetric interaction could manifest itself in an asymmetric way if right-handed currents interact via fields that are much more massive than the ones that couple to left handed currents. The most recent treatment that (a) put Lipmanov's ideas into the framework of a gauge theory, (b) suggested that parity be regarded as a symmetry of the Nambu-Goldstone genus, (c) formulated the principle now known as manifest left-right symmetry and, (d) in the tradition of Lee and Yang, gave a complete analysis of all experiments done to date and thereby showed that this symmetry is not ruled out, is the paper by Bég, Budny, Mohapatra and Sirlin.[18] [There exist many papers which contain some but not all of the features listed above.] The electroweak group, used in this paper, is $U(1)_{b-\ell} \otimes SU(2)_L \otimes SU(2)_R$, b and ℓ being baryon and lepton numbers respectively; a grand unification group which accommodates it is SO(10).

It behooves us, however, to take a look at the other side of the coin; to not do so would be intellectually dishonest.

Our discussion of the shortcomings of conventional theory begins with the observation that theories that are not asymptotically free appear to be afflicted with Landau ghost poles[19] and concomitant paradoxes, and therefore can not be deemed to be consistent as fundamental (as opposed to effective) theories. The prime example is

Quantum Electrodynamics, wherein Landau first noted the possibility of such singularities and where two schools of thought have emerged: the first hopes, but has so far failed to establish, that non-perturbative formulations of QED will be free of Landau poles, or, in other words, that the β-function will develop a non-trivial zero; the second accepts the reality of the paradox and suggests that the logical structure of QED is incomplete. The latter view is the more widely held, and implies the existence of a cut-off momentum, $q_{max.}$ beyond which the theory must break down. To estimate $q_{max.}$ we may use the one-loop expression for the β-function[1] and thereby obtain:

$$\bar{\alpha}(-q^2)^{-1} = \alpha(m^2)^{-1} - \frac{1}{3\pi}\,\ell n(-q^2/m^2) \tag{1.2}$$

where $\bar{\alpha}$ is the running fine structure constant and, if m be chosen to be the electron mass, $\alpha(m^2) \simeq 1/137$. Now physics dictates that $\bar{\alpha}$ be positive semi-definite; hence the inequality

$$\alpha(m^2) \leqq \left[\frac{1}{3\pi}\,\ell n(-q^2/m^2)\right]^{-1} \tag{1.3}$$

must be satisfied for all momenta q in the theory. If the theory is stretched to infinite momenta, and a fundamental theory should have the required capability, we find that $\alpha(m^2) = 0$; this is the Landau paradox. The maximum momentum is therefore given by

$$q^{QED}_{max.} \leqq m_e \exp\left(\frac{3\pi}{2\alpha}\right) \tag{1.4}$$

$$\simeq 5 \times 10^{276}\ \text{GeV} \tag{1.5}$$

The enormous magnitude of $q^{QED}_{max.}$ makes one wonder if one is discussing a point of physics or of philosophy! However, the problem comes back to haunt us in the context of the electroweak synthesis -- which makes use of two asymptotically non-free theories: the U(1) gauge theory and the $\lambda\phi^4$ theory. The difficulties of the latter theory make contact with physics in a more immediate and disturbing way.

With the scalar sector decoupled from fermions and gauge fields, we have a result analogous to Eq. (1.2) for the running quartic self-coupling:

$$\bar{\lambda}(-q^2)^{-1} = \lambda(m_W^2)^{-1} - \frac{3}{4\pi^2}\,\ell n(-q^2/m_W^2). \tag{1.6}$$

where we have defined the renormalized coupling at the W-mass. The magnitude of this coupling is fixed by the mass of the left-over Higgs boson:

$$m_H = m_W(2\lambda \sin^2\Theta_W/\pi\alpha)^{1/2} \tag{1.7}$$

$$\simeq \sqrt{\lambda}\ 355\ \text{GeV, for } \sin^2\Theta_W \simeq 0.2 \qquad (1.8)$$

The statement of the Landau paradox for this theory is $\lambda=0$; this triviality - result for the $\lambda\phi^4$ theory, while not quite proven in the mathematical sense, has been verified through several alternate lines of reasoning and non-perturbative calculational techniques, notably numerical solution of the latticized theory followed by a transition to the continuum limit[20]. Now, the traditional interpretation of the Salam-Weinberg theory rests upon the ability of the scalar sector to generate Goldstone modes, which can be combined with Yang-Mills quanta to yield massive spin-1 bosons. Free field theories, however, can not yield Goldstone bosons.

Pending formulation of a more satisfactory scheme, can one make do by reinterpreting the Salam-Weinberg theory as some sort of effective theory? Unless the Higgs mass is fairly small, the answer is "no". Thus for $m_H \simeq 600$ GeV, the maximum momentum allowed in the theory -- determined by a procedure analogous to that which led to Eq. (1.4) -- is $q_{max.}^{EW} \simeq 800$ GeV. This is within the reach of existing accelerators, far too low for comfort. Unlike the situation with QED, we are not dealing here with merely a point of principle; we have rather a real physical problem which merits careful scrutiny.

A tentative theoretical scenario, in which the domain of validity of the standard model is considerably enlarged, will be discussed in these lectures. That it can not but lead to relatively small values of the Higgs mass is to be expected; that it can not be the last word will be obvious.

Attempts at a definitive resolution of the difficulties of conventional theory, which do not appeal to any radical or unfamiliar concepts, invoke two strategies that need not be unrelated. The first simply postulates that "new physics" will manifest itself at energies of the order of a TeV; the second goes after the root of the problem and tries to eliminate elementary scalars[21]. If one uses the hypercolor/technicolor construct[22] to remove scalars from the Lagrangian, one naturally finds reasonably-well-specified new physics in the TeV regime; herein lies the principal appeal of a subject which has been beset from the beginning with theoretical problems that seem to be well-nigh insoluble. It should be noted though that hypercolor can not resolve all difficulties of the standard model; problems stemming from the U(1) factor in the gauge group will remain; the concept is more likely to be useful in the context of grand unification based on a simple (or semi-simple) group. In these lectures we shall go over the subject of hypercolor in some detail, and weigh it in the balance; we shall find that, at least in its present state, it must be deemed to be wanting.

Returning to canonical theory, the difficulties stemming from the triviality of dynamical systems consisting of self-coupled scalars take a turn for the worse in the context of grand unification based on a semi-simple group, such as SU(5). Even the provisional scenario used for the SU(3) ⊗ SU(2) ⊗ U(1) based standard model is no longer available. Recognizing that SU(5) did render QED consistent, it may not be imprudent to conjecture that the problems of SU(5) itself may be resolved by the next logical step in the unification program: a synthesis with gravitation. Before this can be achieved, however, some means must be found for taming the singularities of gravitational theory; within the framework of conventional field theory in (1+3)-dimensional space, this may not be possible.

For pure Einstein gravity[20], a remarkable cancellation of infinities does occur at the one-loop level in (1+3)-space; however, at the two-loop level the unrenormalizable infinities return; in the presence of matter, they manifest themselves already at the one-loop level. One may attempt to soften things by generating gravity, as part of supergravity, through local super-symmetry. Pure supergravity theories -- that is to say, theories in which the spectrum consists entirely of gravity supermultiplets -- exhibit one-loop finiteness, just as pure gravity does; furthermore, unlike the situation in pure gravity, the finiteness persists (at least) upto the two-loop level; this circumstance led to much activity in the recent past, and kindled the hope that a sensible description of the coupled matter-gravity system may be in hand. However, the matter spectrum that is part of the standard model can not be fitted into the procrustean bed of gravity supermultiplets; even one loop finiteness is thus not attained in the real world.

We seem to be led, willy-nilly, to the conclusion that in the presence of gravity, grand unified and electroweak theories make less sense than they did on their own, that the cut-off dependence that mars the canonical formulations of these theories is exacerbated not removed. The moral we choose to draw is that one has to go outside the framework of local field theory in (1+3)-space to describe quantum gravity; once an ultra-violet convergent theory of gravity is available, one may proceed with some optimism to complete the logic of grand unification. Very recent developments in the superstring theory of Michael Green and John Schwarz[23] -- which is best formulated in ten dimensions, with the extra six compactified at a scale irrelevant to low energy physics -- give cause for hope.

The problems discussed above change completely in the new context; instead of ascending to the Planck mass one has to avoid the pitfalls in descending from the Planck mass to 100 GeV, the characteristic scale of electroweak physics. (Fig. 1.1).

I trust that you will invite superstring experts to the next "Escuela Mexicana...", to give you the solutions to the problems that I shall pose in these lectures.

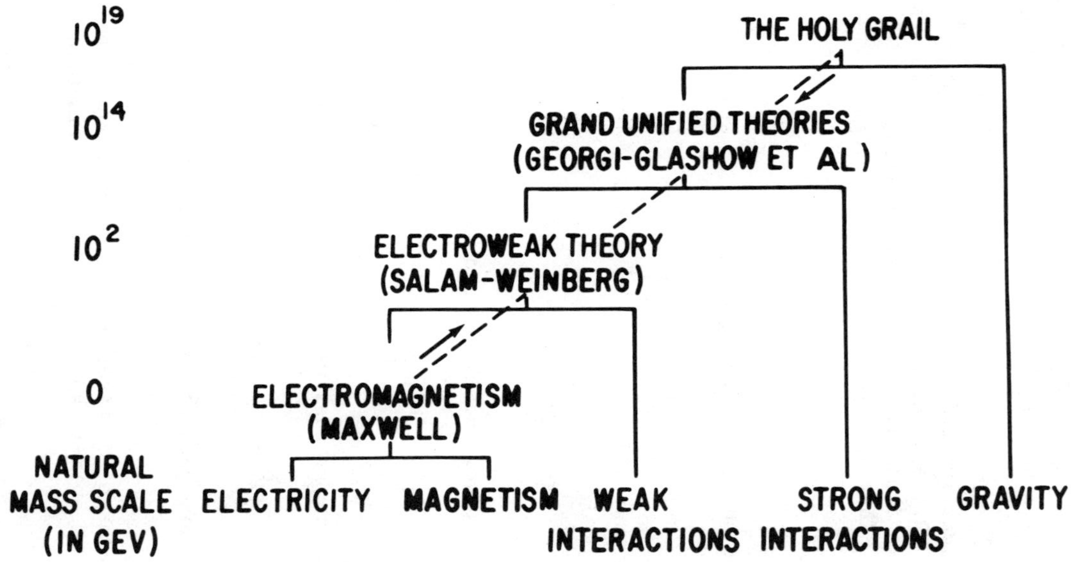

Fig. 1.1 Progressive Unification, circa 1985 C.E. Ascending: conventional theory. Attempting descent: superstring theory.

2. CANONICAL DESCRIPTION OF ELECTROWEAK AND STRONG INTERACTIONS

2.1 Basic Principles of Gauge Theories

The gauge-theoretic formulation of elementary particle physics is predicated, at this time, on the following postulates:

i) All matter is composed of two species of spin 1/2 fermions, "Quarks and Leptons", and bosons which mediate the interactions between the fermions.

ii) All interactions are generated via gauge principles.[1-7]

For instance in QED the electromagnetic interaction is achieved when the Lagrangian is demanded to be invariant under the U(1) local abelian gauge group. That is

$$\mathcal{L}_o = \bar{\psi}(i\gamma_\mu \partial^\mu + m)\psi \tag{2.1}$$

can be made invariant under the transformation

$$\psi(x) \rightarrow e^{i\alpha(x)}\psi(x) \tag{2.2}$$

if the term $i\bar{\psi}\gamma_\mu\partial^\mu\psi$ is replaced by $i\bar{\psi}\gamma_\mu D^\mu\psi$ where

$$D^\mu = \partial^\mu - ie_\psi A^\mu \tag{2.3}$$

is the covariant derivative, and A^μ transforms as

$$A^\mu \rightarrow A^\mu + \frac{1}{e}\,\partial_\mu\alpha(x) \tag{2.4}$$

For non-abelian gauge groups, we have more generally

$$D^\mu = \partial^\mu - iT_\alpha A^{\mu\alpha} \tag{2.5}$$

T_α being the matrix representation of the gauge group in the space of the ψ's.

iii) To every unbroken generator of the gauge group corresponds a massless gauge field of spin-1 (boson with two* polarization degrees of freedom: $\frac{\vec{p}\cdot\vec{s}}{p} = \pm 1$).

*In empty space, far from external sources.

iv) If physics requires that a generator be broken (i.e. the corresponding gauge field to be massive), then its breaking occurs via the Higgs mechanism.

The main steps in the Higgs mechanism can be summarized:

a) The Higgs sector is adjusted to produce a tachyon mode ($\mu^2<0$). In the Weinberg-Salam model

$$V(\phi) \equiv \mu^2\phi\phi^* + \lambda(\phi\phi^*)^2, \quad \lambda > 0 \tag{2.6}$$

and the minimum of V occurs at $|\phi| = \frac{v}{\sqrt{2}} = (-\mu^2/2\lambda)^{1/2}$.

b) Instability leads to spontaneous symmetry breaking, which implies the existence of Goldstone bosons.

c) The Goldstone bosons plus massless gauge fields yield massive spin-1 bosons.

The restriction that the symmetry is broken by the Higgs mechanism is connected with the requirement that one needs to have a renormalizable theory. (The Higgs mechanism secures an ultra-violet-behavior of Green's functions identical to that in the unbroken theory; hence, if the symmetric theory is renormalizable, so is the theory with broken symmetry).

2.2 Flavor and Color Spaces

The electroweak interactions stem from gauge principles in flavor space, while the strong interactions stem from gauging the color degree of freedom. The $SU(2)_L \otimes U(1)_Y$ and $SU(3)_C$ symmetry groups, of electroweak and strong interactions respectively, may be realized as commuting groups because of the orthogonality between flavor and color spaces.[2]

The quarks u, d, c, s, t and b transform as triplets under $SU(3)_C$, the color group, while the leptons ν_e, e, ν_μ, μ, ν_τ, τ are singlets (invariant) under this group.

2.3 The $SU(3)_C \otimes SU(2)_L \otimes U(1)_Y$ Based Standard Model[2]

Particle spectrum, masses and mixings.

Because of its success in describing low energy phenomenology and its relative economy in the number of fundamental fields, the electroweak model constructed on the basis of the spontaneously broken $SU(2)_L \otimes U(1)_Y$ gauge group[3,4] has been incorporated into what is known as the "standard-model". As this model has been widely surveyed[5], we shall limit ourselves to a rather brief description of its general features and properties mainly to set the stage for further discussions.

At present it is widely believed that QCD, i.e. the unbroken gauge theory based on the color group $SU(3)_C$, is the most promising candidate to describe the strong interactions[6]. Thus the electroweak and strong interactions at low energies are deemed to be governed by a gauge theory based on the $SU(3)_C \otimes SU(2)_L \otimes U(1)$ group.

To further specify the theory, one must describe its representation content: left-handed quarks and leptons behave as doublets under $SU(2)_L$ (hence the subscript L), while their right-handed counterparts are singlets; quarks, endowed also with the color degree of freedom, transform as triplets of $SU(3)_C$. Four intermediate bosons, $W^{\pm}$, Z^o and γ, mediate the electroweak interactions; eight massless gluons transforming according to the octet representation of $SU(3)_C$ are associated with the strong interactions.

In the minimal scheme there are two complex Higgs scalars which are doublets of $SU(2)_L$ and singlets of $SU(3)_C$

$$\phi = \begin{pmatrix} \phi^{(+)} \\ \dfrac{\phi_1^o - i\phi_2^o}{\sqrt{2}} \end{pmatrix} \tag{2.7}$$

where $\phi^{(+)}$ is complex and ϕ_1^o, ϕ_2^o are real.

One of the neutral Higgs scalars (ϕ_1^o) develops a non-vanishing vacuum-expectation-value (V.E.V). As a consequence the theory is spontaneously broken: three of the Higgs scalars ($\phi^{(+)}$, ϕ_2^o) are absorbed in a redefinition of the gauge bosons, three of which ($W^{\pm}$, Z) become massive, while the fourth (the photon A) is associated with the unbroken $U(1)_Q$ group and remains massless. (ϕ_1^o can be written as $\phi_1^o = \langle\phi_1^o\rangle + \phi$, where ϕ is left over as a physical Higgs.) This is a particular example of the Higgs-Kibble mechanism.

Detailed study shows that the mass matrix eigenstates Z and A are orthogonal combinations of the neutral gauge fields W_μ^3 and B_μ associated with the $SU(2)_L$ and U(1) subgroups:

$$W_\mu^3 = \cos\Theta_w Z_\mu + \sin\Theta_w A_\mu \tag{2.8}$$

$$B_\mu = \cos\Theta_w A_\mu - \sin\Theta_w Z_\mu \tag{2.9}$$

The weak interaction angle Θ_w is given by

$$\tan\Theta_w = g'/g \tag{2.10}$$

where g and g' are the gauge couplings of $SU(2)_L$ and U(1), respectively. One further finds (the $\Delta I_{wk} = 1/2$ rule):

$$m_z \cos\Theta_w = m_w \tag{2.11}$$

a relation that still holds true if, for example, additional doublets and singlets of Higgs scalars are introduced. More generally[7]

$$\rho \equiv \frac{m_W^2}{m_Z^2 \cos^2\Theta_W} = \frac{\sum_i [I^{(i)}(I^{(i)}+1) - I_3^{(i)2}]\lambda_{(i)}^2}{2\sum_i I_3^{(i)2}\lambda_{(i)}^2} \qquad (2.12)$$

where the sums are over the representations of scalar fields with weak isospin $I^{(i)}$; $I_3^{(i)}$ and $\lambda_{(i)}$ are the third component of weak isospin and the vacuum expectation value of the neutral member of the ith multiplet.

At the tree level there are two further important relations involving the coupling constants: the proton electric charge e and the muon decay coupling constant G_μ are related to the basic parameters of electroweak theory by means of the expressions

$$e = g \sin \Theta_W \qquad (2.13)$$

$$G_\mu/\sqrt{2} = g^2/8m_W^2 \qquad (2.14)$$

To generate ordinary fermion masses, quarks and leptons are deemed to have gauge invariant Yukawa interactions with the Higgs scalars. Mass terms for the leptons and quarks are generated through those couplings as a consequence of the non-vanishing expectation value of the neutral Higgs field. The mass matrices for the quarks and leptons induced in this fashion are, in general non-diagonal, but they can be diagonalized by suitable unitary transformations on the left-handed and right-handed fields. When the fields coupled to the gauge bosons are expressed in terms of these mass eigenstates it is found that the $SU(2)_L$ doublets can be written as

$$\begin{pmatrix}\nu_e\\ e^-\end{pmatrix}_L, \begin{pmatrix}\nu_\mu\\ \mu^-\end{pmatrix}_L, \begin{pmatrix}\nu_\tau\\ \tau^-\end{pmatrix}_L \quad \text{(leptons)} \qquad (2.15)$$

$$\begin{pmatrix}u\\ d'\end{pmatrix}_L, \begin{pmatrix}c\\ s'\end{pmatrix}_L, \begin{pmatrix}t\\ b'\end{pmatrix}_L \quad \text{(quarks)} \qquad (2.16)$$

where the primes in d', s', b' are meant to indicate that they are linear combinations of the mass eigenstates d, s, b with the two sets of fields related by the Kobayashi-Maskawa matrix[8].

$$\begin{pmatrix}d'\\ s'\\ b'\end{pmatrix} = \begin{pmatrix} c_1 & s_1c_2 & s_1s_2 \\ -s_1c_3 & c_1c_2c_3 - s_2s_3e^{-i\delta} & c_1s_2c_3 + c_2s_3e^{-i\delta} \\ -s_1s_3 & c_1c_2c_3 + s_2c_3e^{-i\delta} & c_1s_2s_3 - c_2c_3e^{-i\delta}\end{pmatrix}\begin{pmatrix}d\\ s\\ b\end{pmatrix} \qquad (2.17)$$

$$c_i = \cos\Theta_i$$

$$s_i = \sin\Theta_i \qquad (2.18)$$

where Θ_1, Θ_2, Θ_3, δ are mixing parameters that may be determined experimentally[9].

In the lepton sector we have assumed the neutrinos to be massless; therefore any possible mixing of the mass eigenstates can be removed by a suitable redefinition of the ν fields. If the neutrinos are massive and there are no global conservation laws of electron, muon and τ numbers, there will in general exist mixings in Eq. (2.15) similar to those occurring in Eq. (2.16).

2.4 Symmetry of the Higgs Sector

We know that in the standard model, the Higgs sector (Eqs. (2.6) and (2.7)) has an $O_4 \equiv SU(2)_L \otimes S(2)_R$ global symmetry. When $<\phi> = \frac{<\phi_1^o>}{\sqrt{2}} \neq 0$, this O_4 is spontaneously broken to $SU(2)_{L+R}$. To consistently couple the Higgs sector to the gauged weak-interaction sector, the global symmetry of the Higgs sector need only be $SU(2)_L \otimes U(1)$. The extra symmetry, the so-called custodial isospin, inescapably leads to the mass relation $M_w^2/M_z^2\cos^2\Theta_w = 1$ to lowest order in the gauge coupling and all orders in the scalar self-coupling. Thus at the tree level $\rho = 1$, but $m_u \neq m_d$. Small deviations from $\rho = 1$ will arise from higher-order gauge interactions which do not respect $SU(2)_{L+R}$.

At one loop, $\Delta\rho(\equiv \rho-1)$ receives contributions from fermion loops inserted in the gauge-boson propagator (Fig. 2.1). The result is given by[10,11]

$$\Delta\rho = \xi \frac{1}{16\pi} \frac{\alpha_w}{M_w^2}\left[\frac{2m_u^2 m_d^2}{m_u^2-m_d^2} \ln \frac{m_d^2}{m_u^2} + m_u^2 + m_d^2\right] \qquad (2.19)$$

where m_u and m_d are the masses of the individual fermions within a doublet and ξ is a color factor (which is 1 for leptons and 3 for quarks).

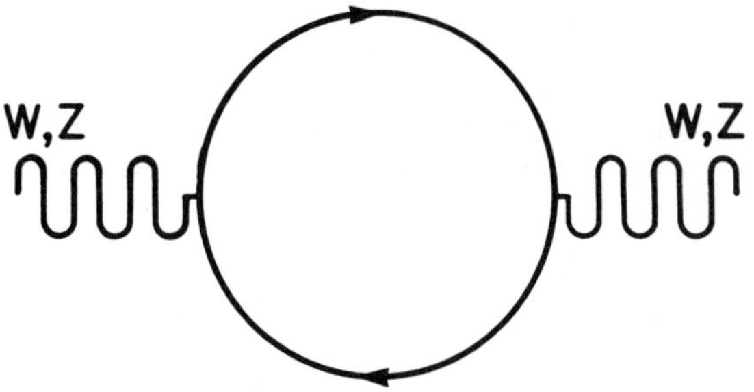

Fig. 2.1 Fermion-loop correction to the gauge-boson self-energy diagram.

2.5 Status of QCD

The color quantum number which was postulated to resolve the spin-statistics problem also received strong confirmation from the $\pi^o \to 2\gamma$ decay[12] and attempts to understand the ratio $R = \sigma(e^+e^- \to \text{hadrons})/\sigma(e^+e^- \to \mu^+\mu^-)$ from a quark picture.

We now know that the most serious candidate to describe the strong interactions (symmetry properties, quarks, colors, asymptotic freedom, etc.) is Quantum Chromodynamics (QCD), a non-abelian gauge theory based on the unbroken gauge group $SU(3)_C$ of color. All experimental results obtained to date are consistent with QCD; none contradicts it.

The infra-red singularities in QCD are much more severe than in QED; in perturbation theory, for example, there are no concellations analogous to those that would occur in QED between the following diagrams:

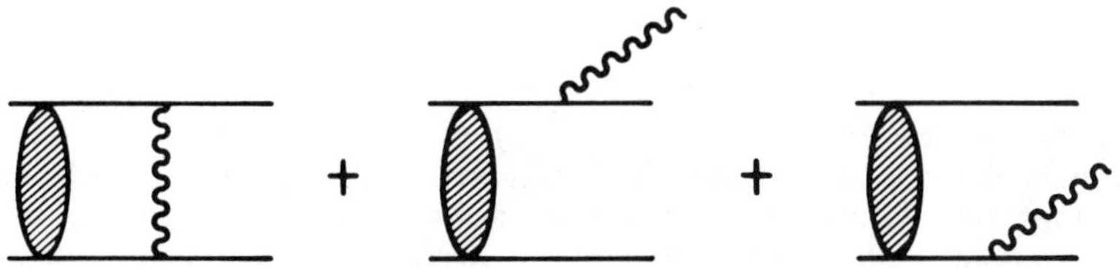

Fig. 2.2 Radiative corrections to quark-quark scattering.

This problem is however resolved if the theory:
a) chooses the Higgs phase or b) chooses confinement. The ultraviolet singularities which arise at the one loop level can be absorbed into the bare parameters; this argument can be extended to multiloop graphs as well, and thus permits us to establish that QCD is a renormalizable theory.

2.6 Status of Standard QFD

By standard QFD we shall mean the description of electroweak interactions between quarks and leptons based on the gauge group $SU(2)_L \otimes U(1)$ and within the framework of the canonical methodology (i.e. symmetry-breaking a la Higgs-Kibble, with the help of elementary spin-0 fields, to generate masses for gauge fields which couple to broken generators). Commitment to a specific representation content is part of our definition of standard QFD: It is thus understood that the four gauge fields, $W^{\pm}$, Z and γ, are accompanied by one physical Higgs field ϕ and that there are three generations of quarks and leptons.

As we explained in Section 2.3 there is a mismatch between the quark fields of definite current mass and the fields which transform irreducibly under the weak isospin group, the latter being the ones which occur in the Lagrangian at the first stage of symmetry breaking. Diagonalization of the quark mass-matrix is thus necessary, and it introduces four parameters into the theory: three Cabibbo-like angles and one CP-violating phase. (The story is repeated in the leptonic sector if neutrinos have the attribute of mass.) Standard QFD is thus a theory in which all the relevant experimental results can be fitted with the following 17 parameters: six quark masses, three lepton masses, two gauge field masses, the Higgs field mass, four "mixing" angles, the fine structure constant α and a Θ-like parameter (see Section 2.7.1. below); while these add up to 18, one of the masses may, without loss of generality, be set equal to unity.

2.7 Interface of QFD and QCD

The union of standard QFD and QCD, that is to say, the theory based on the gauge group $SU(3)_C \otimes SU(2) \otimes U(1)$ with the above representation content constitutes the standard model. Within this model the application of current algebra to weak decays has been by and large successful. However there are a number of important and fundamental questions about the nature of the electroweak interactions that remain unanswered (for instance: What is the origin of CP-violation?)

2.7.1. Strong CP - problem

The strong CP-problem refers to the potentially dangerous leakage of CP and P non-invariance effects from the QFD sector to the

QCD sector in the presence of instanton-like configurations; it is the price one pays for resulution of the $U_A(1)$ problem in QCD along the lines suggested by 't Hooft[13]; the problem is therefore also known as "the new $U_A(1)$ problem". This leakage occurs if the transformation which diagonalizes the quark-mass matrix contains a factor belonging to the group $U_A(1)$; its effect is to generate in the QCD Lagrangian an additional term[2]

$$\delta L = \frac{\Theta_{QFD}}{46\pi^2} \text{Tr} \cdot \varepsilon^{\mu\nu\rho\lambda} G_{\mu\nu} G_{\rho\lambda} \tag{2.20}$$

where G is the covariant curl of the color-gluon-field matrix* and

$$\Theta_{QFD} = i \, \ell n(\det M / \det M^{+}) \tag{2.21}$$

M being the R → L mass matrix.

It is the simultaneous violation of P and CP by δL which makes the strong CP-problem a problem; this feature in a flavor-diagonal part of the Lagrangian permits it to contribute to the electric dipole moment of the neutron[14,15]; the challenge is to reconcile the magnitude of the contribution with the latest Ramsey limit.**

Eq. (2.20) is a consequence of the Bell-Jackiw-Adler anomaly[17] for the color-singlet flavor-singlet axial current in QCD. The anomaly is a short distance phenomenon[18], the anomalous divergence can be evaluated in perturbation theory; yet Eq. (2.20) is non-trivial only in the presence of non-perturbative long-range effects such as come into play in the presence of instantons.[19] This is so because δL is expressible as a 4-divergence and does not contribute to amplitudes in any finite order of perturbation theory; in the absence of instanton-like configurations it does not in fact contribute to the action. Quite obviously we are dealing here with an ultraviolet infra-red connection of the sort which often surfaces in non-abelian gauge theories.

If Θ_{QFD} is infinite, or finite but unacceptably large, one is obliged to work with a OCD Lagrangian containing a counter term of the same form as δL such that

$$\Theta_{effective} = \Theta_{QCD} + \Theta_{QFD} \tag{2.22}$$

may be chosen to be of an admissible order of magnitude. It is widely recognized, however, that controlling strong P and CP non-invariance effects in this manner is not a terribly enlightening procedure.

* The QCD coupling constant has been absorbed in G.

** The most recent limit is $d_n < e.3 \times 10^{-24}$ cm (See Ref. 16).

3. HIGGS MASS IN SALAM-WEINBERG THEORY

I shall next discuss the consistency of the canonically realized Salam-Weinberg theory[1] under the premise that the pure $\lambda\phi^4$ theory is a trivial field theory in four dimensions[2]. Recall that by "canonically realized" one means (a) that the gauge-group underlying electroweak and strong interactions is $U(1) \otimes SU(2)_L \otimes SU(3)_C$ and (b) that the Higgs mechanism which reduces the symmetry of the electroweak interactions to $U(1)_Q$, the electromagnetic gauge group, is triggered with the help of the usual elementary spin-0 fields, to wit: four real fields that can be combined into a complex doublet, so that three are devoured to produce massive $W^{\pm}$ and Z and one manifests itself as a physical particle even[3] under CP. As noted in Section 1, this construct faces serious problems of consistency if the $\lambda\phi^4$ theory is trivial; the usual interpretation, wherein one presumes that the ϕ-sector by itself produces the Goldstone bosons needed for the Higgs mechanism, must be abandoned. We may, however, demand that the coupled Higgs- gauge -field system be non-trivial; for a wide range of quark masses, this leads to an upper bound of about 125 GeV for the mass of the physical Higgs boson[4].

3.1 Some Rigorous Results

Before I explain how the Higgs mass can be bounded, let me review the status of the input-premise: the triviality of the $\lambda\phi^4$ theory in the real world. I should emphasize that, strictly speaking, we are dealing here with an article of faith; a rigorous demonstration of the triviality of the continuum theory is not yet in hand. Nonetheless, one may predicate one's faith on some results that have been derived with a measure of rigor; these are listed below. An essential ingredient in many of them is the postulate that one may move from Euclidean space formulation of the theory to Minkowski space (real time) without let or hiderance; in technical jargon, the solutions of the field equations are supposed to be in accord with the Osterwalder-Schrader axioms.[5]

(i) Triviality cannot be proven in any finite order of perturbation theory.

This almost obvious result can be put on a firm basis, using the work of Glimm and Jaffe.[6]

(ii) The renormalized coupling constant lies in a bounded interval.[7]

$$0 \leq \lambda_{ren} < \lambda_{max} \text{ for } d \leq 4, \tag{3.1}$$

where d is the dimensionality of the space in which the theory is defined.

A necessary condition for triviality is thus satisfied.

(iii) For the theory in the symmetric phase, triviality for $d>4$ has been established by Aizenman[8] and Frohlich.[9]

(iv) For d = 4, Frohlich[9] has noted that triviality can be established if Z_3 - the wave-function renormalization constant-vanishes.

(v) The continuum limit of the lattice theory is trivial, or consistent with triviality, in all existing calculations.[10] When other nonperturbative calculational techniques are available, notably the 1/N expansion for the O(N) - symmetric theory, triviality again follows in the limit of infinite cutoff.[11]

Recently Bég and Furlong[12] have added one more result to the above list:

(vi) The nonrelativistic limit of the theory is trivial in 1+3 dimensions. The collision matrix vanishes and so does the renormalized coupling; the S matrix thus reduces to the unit matrix.

Paragraph (v) summarizes what appear to be, at this time, the most compelling reasons for believing that the theory is indeed trivial. In the following we[4] accept this result; our apologia is that if it turns out that there is no disease, our cure -- while deprived of a sound raison d'etre will do no harm.

3.2 Heuristic Considerations

To get some insight into the nature of the problem, and the manner in which it can lead to an upper bound for the Higgs mass, let me sketch a heuristic non-proof of triviality based on Eq. (1.6).

The renormalization group invariant or running coupling constant, $\overline{\lambda}$, may be written at the one-loop level[13] as

$$\overline{\lambda}(t)^{-1} = \lambda^{-1}_{Ren.} - (3/2\pi^2) \cdot t \tag{3.2}$$

where $2t = \ln(p^2/m_w^2)$, p is the momentum variable customarily used in renormalization group calculations and m_W, the mass of the W-boson, has been used to introduce a mass scale.

Now stability requires that $\overline{\lambda}(t)$ be positive semi-definite; hence $p_{max.}$, the largest admissible momentum in the theory, must be such that

$$\lambda_{Ren.} \leq (2\pi^2/3)\left[\ln\ (p_{max.}/m_w)\right]^{-1} \tag{3.3}$$

If $p_{max.} \to \infty$, $\lambda_{Ren.} \to 0$ and we get a trivial theory.

If $\lambda_{Ren.} = 0$, the Salam-Weinberg theory collapses. There are no would be Goldstone bosons in any gauge, there is no Higgs mechanism and $M_W = M_Z = 0$, contrary to observation: $M_W \simeq 81$ GeV, $M_Z \simeq 92$ GeV (CERN, UA1, UA2).

If non-triviality is achieved with some definite mechanism corresponding to a finite cut-off, Eq. (3.3) can be converted into an upper bound for the Higgs mass via[1]

$$M_H^2 = M_W^2\ 8\lambda_{Ren.} \sin^2\Theta_w/e^2 \tag{3.4}$$

where Θ_w is the electroweak angle and e is the electric charge.

Observe, from Eqs. (3.3) and (3.4), that $p_{max} = 10^{38}$GeV corresponds to $M_H < 100$ GeV, $p_{max} = 10^{19}$ GeV to $M_H < 150$ GeV.

Needless to say, the introduction of cut-offs that can not be pushed to infinity destroys the mathematical structure of a field theory, reducing it to a kind of phenomenology. While introduction of ad hoc cut-offs is not a defensible proposition, effective cut-offs can arise naturally in grand unified theories or theories which bring gravity into the picture via, for example, embedding of electroweak theory in a local supersymmetric theory. In such theories, however, the triviality problem acquires a new dimension; I have nothing to say about it at this time. Let us return, therefore, to our original mission: investigation of the consistency of the canonically realized theory.

3.3 A Framework for Resolution

The framework which we[4] (Constantinos Panagiotakopoulos, Alberto Sirlin and I) offer for resolution of the conundrum may be outlined as follows:

(i) We propose that the coupled Higgs -- gauge-field sector is non-trivial.

(ii) To uphold (i), we require that y $[\equiv\bar{\lambda}(t)/\bar{g}_1(t)^2]$ does not diverge or become unduly large for "large" t.

Here g_1 is the coupling constant for the U(1) factor in the strong-electroweak group (with g_2 and g_3 defined in an analogous way); the precise meaning of "large" will become clear in the following.

Observe that (ii) may be deemed to be a necessary condition for consistency; if it is not satisfied, the ϕ-sector would decouple from the rest of the Lagrangian; arguments for the triviality of this sector would then go through.

(iii) To implement (ii) we demand that y be driven to an ultra-violet stable fixed point of the renormalization group.

That y must indeed be of reasonable magnitude if it is to make it to a fixed point will be demonstrated below. As noted in Ref. 13, $\bar{\lambda}(t)/\bar{g}_2(t)^2$, the other ratio that one may be tempted to examine, can not go to a fixed point.

3.4 Renormalization Group Analysis and Results

To establish our notation and normalization, we display explicitly the Lagrangian density corresponding to the Higgs sector.[14]

$$L_\phi = (\partial_\mu\phi^\dagger - \frac{i}{2} g_1\phi^\dagger B_\mu - \frac{i}{2} g_2\phi^\dagger\vec{\tau}\cdot\vec{A}_\mu)(\partial^\mu\phi + \frac{i}{2} g_1 B^\mu\phi + \frac{i}{2} g_2\vec{\tau}\cdot\vec{A}^\mu\phi)$$
$$-\mu^2\phi^\dagger\phi - \lambda(\phi^\dagger\phi)^2 + G\bar{\psi}_L\tilde{\phi}\psi_R + \ldots \qquad (3.5)$$

Here B and A are the U(1) and $SU(2)_L$ gauge fields, respectively; the τ are the Pauli matrices; $\mu^2<0$, for spontaneous symmetry breaking; $\lambda > 0$, for stability; and the Higgs field is assigned hypercharge $Y = +1$. The ψ's are Fermi fields; we have displayed only one of the many possible Fermi Higgs Yukawa couplings [$\tilde{\phi} \equiv i\tau_2(\phi^\dagger)^T$]; the dots indicate the others.

At the one-loop level, the renormalization-group equations for the gauge couplings are

$$\frac{d\bar{g}_i}{dt} = \varepsilon_i b_i \bar{g}_i^3/16\pi^2, \tag{3.6}$$

where $i(=1,2,3)$ labels the gauge group, $b_i>0$, $\varepsilon_1 = +1$, and $\varepsilon_2=\varepsilon_3= -1$. Before we write down the one-loop β functions for the other couplings, we introduce variables

$$x \equiv \bar{g}_2^2/\bar{g}_1^2;\ \zeta \equiv \ell n[\ \bar{g}_1^2/\bar{g}_1^2(t = 0)];$$

$$z \equiv \bar{G}^2/\bar{g}_1^2;\ u \equiv \bar{g}_3^2/\bar{g}_1^2\ ,$$

where G is now identified as the Yukawa coupling associated with the top quark.

Note that x, ζ, and u are simply related by virtue of Eq. (3.6). We have

$$u(x) = (b_1/b_3)Cb_2x/\ [b_1+xb_2(1-C)], \tag{3.7a}$$

$$x = \frac{(b_1/b_2)C'\exp(-\zeta)}{1-C'\exp(-\zeta)}\ , \tag{3.7b}$$

where C and C' can be determined from our knowledge of the coupling constants at $t = 0$, corresponding to $\zeta = 0$ and $x = 1/\tan^2\Theta_W$, Θ_W being the electroweak angle. (The domain of ζ, $0 < \zeta < \infty$, corresponds to the interval $\tan^{-2}\Theta_W > x > 0$.) Note further that at the tree level,

$$y(t = 0) = m_H^2/(8m_W^2\tan^2\Theta_W)\ , \tag{3.8a}$$

$$z(t = 0) = m_t^2/(2m_W^2\tan^2\Theta_W)\ . \tag{3.8b}$$

The renormalization-group equations for the quartic and Yukawa couplings[15] can be put in the form

$$\frac{dy}{dx} = -\ \frac{192y^2-8y(3+2b_1+9x-12z)+3(1+2x+3x^2)-48z^2}{16x(b_1+b_2x)} \tag{3.9a}$$

$$\frac{dz}{dx} = - z\,[9z-2b_1-16u(x)]/2x(b_1+b_2x). \tag{3.9b}$$

[Some electroweak contributions, small in comparison with $2b_1+16u(x)$, have been neglected in Eq. (3.9b).]

Both equations are of the Riccati type.[16] The second can be solved exactly; using ζ as the independent variable, we have

$$z(\zeta) = \frac{z(0)\exp[\chi(\zeta)]}{1+z(0)(-9/2b_1)\int_0^\zeta \exp[\chi(\zeta)]d\zeta}, \tag{3.10}$$

where $\chi(\zeta) = -b_1^{-1} \int_0^\zeta (b_1+8u)d\zeta$.

To avoid the singularity - in other words, have a solution that tends to the fixed point $z = 0$ as $\zeta \to \infty$ - that would otherwise develop in $z(\zeta)$, $z(0)$ must be bounded from above; this leads to the upper bound M($\cong$168 GeV) on the mass of the top quark.

Equation (3.9a) can be solved analytically for small x and z; this is adequate, however, only to show (in a constructive way!) that there exists a nonsingular solution which is driven to the fixed point

$$y^* = [1 + \tfrac{2}{3}b_1 - \{(1 + \tfrac{2}{3}b_1)^2-4\}^{1/2}]/16 \tag{3.11}$$

as $x \to 0$. To proceed further, it is necessary to integrate Eqs. (3.9a) and (3.9b) numerically; this was done with use of a computer program based on the Runge-Kutta method,[16] with the following values of the input parameters:

$$b_1 = \frac{20}{9}n_g + \frac{1}{6} = 6.83333; \tag{3.12a}$$

$$b_2 = \frac{22}{3} - \frac{4}{3}n_g - \frac{1}{6} = 3.16666, \tag{3.12b}$$

$$b_3 = 11 - \frac{4}{3}n_g = 7, \tag{3.12c}$$

$$x_{init} \equiv x_o = 3.54545, \tag{3.12d}$$

$$u_{init} \equiv u_o = 10.0987. \tag{3.12e}$$

The numerical values correspond to n_g (number of generations) = 3, $\sin^2\Theta_W = 0.22$, $\Lambda_{\overline{MS}} = 0.1$ GeV, and initialization at momentum $m_W \approx 81$ GeV. ($\Lambda_{\overline{MS}}$ is the usual QCD parameter.)

The computer generated solutions of Eqs. (3.9a,b) are consistent with the basic fact that if the theory is to be in the domain of attraction of an ultra-violet stable fixed point, its parameters can not be arbitrary; in particular, the Higgs mass must lie in a bounded

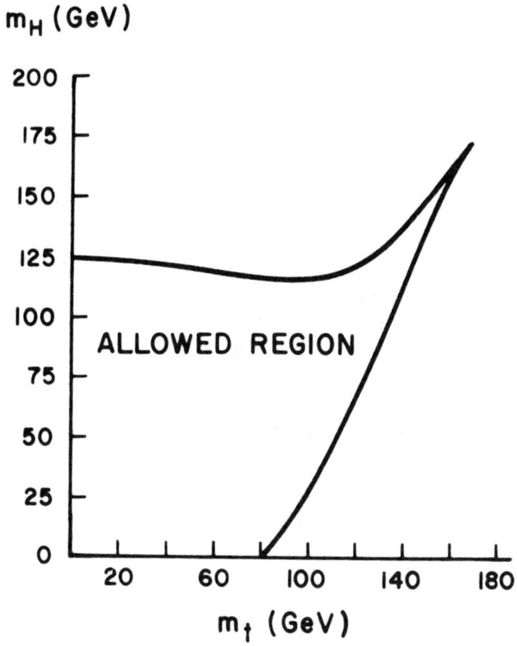

Fig. 3.1 Upper and lower bounds on the Higgs-boson mass plotted as functions of the top-quark mass. [The lower bound of Weinberg and Linde ($m_H \geqq 7$ GeV; see for example Ref. 1) is not depicted.]

interval. We distinguish three cases:

(a) If m_t, the mass of the top quark (or any other heavy quark), is less than a determinable value, say m^*, then

$$0 < y(t = 0) \leqq Y_{max}(m_t), (m_t \leqq m^*). \tag{3.13}$$

(b) If m_t exceeds m^*, there is also a nontrivial lower bound for y_{init} which arises from the requirement that y remain positive definite for all t:

$$Y_{min}(m_t) \leqq y(t = 0) \leqq Y_{max}(m_t), \quad (m_t > m^*). \tag{3.14}$$

(c) If m_t equals M, the maximal value allowed in our formulation, the upper and lower bounds on y(t = 0) coalesce into one; the Higgs mass is then determined rather than bounded.

We are now in a position to state our numerical results for m_H, the Higgs-boson mass, obtained in the manner explicated above.

Case (a): For $m_t < m^* \simeq 80$ GeV, we find that $(m_H/m_W)^2 \leqq 2.376$, corresponding to $m_H \leqq 125$ GeV. It is obvious that this is the most interesting case, the bound being the lowest of the many upper bounds on Higgs mass that have been published in the literature.[17]

Case (b): For m_t>80 GeV, the bounds vary quite sharply with m_t. Thus, 65 GeV < m_H < 122 GeV for m_t = 120 GeV, while 140 GeV<m_H<148 GeV for m_t = 150 GeV.

Case (c): For m_t = 168 GeV, the largest possible value, $m_H \simeq$ 175 GeV.

The bounds on m_H are plotted as functions of m_t in Fig. (3.1).

3.5 Extensions of the Scenario

The results stated above have been extended and generalized by several authors. Babu and Ma[18] have shown that the analysis is stable under inclusion of two-loop contributions to the β-functions; for the bound in Case (a) they found

$$m_H < 130 \text{ GeV} \quad \text{(1 loop and 2 loops)} \tag{3.15}$$

Extension to N Higgs doublets was first carried out by Bovier and Wyler[19]; they showed that, for $m_t \simeq 0$, the mass of the lightest Higgs particle satisfies

$$m_H < \frac{125}{\sqrt{N}} \text{ GeV} \tag{3.16}$$

For N = 2, Eq. (3.16) agrees with an independent analysis by Babu and Ma[20]. These authors have also extended the above discussion to the case of 4 fermion generations[20].

3.6 Questions of Principle

Before we proceed to a critique of our analysis let us note that what was promised at the outset has been delivered. First, the requirement of a reasonable magnitude for y is indeed met if it goes to a fixed point. Thus for m_t = 31 GeV, $y(x_o)$ = 1.02, we find that $0.027 < y(x) < 1.02$ for $5 \times 10^{-5} < x < x_o$ (= 3.54545). This range of y may be compared to the value at the fixed point: $y = y^* \simeq 0.023$. Second, if $y(x_o)$ is such that the fixed point is not attained, a catastrophic blow-up can in fact occur; for example, with $y(x_o)$=1.10 we find that $y \sim 4\times10^8$ for $x \sim 0.16$.

There is a shortcoming in our formulation which, paradoxically enough, permits us to give a precise meaning to the words "large t". For g_1, one has a singularity similar to the Landau-Pomeranchuk singularity[21] in pure QED. Crossing this singularity would take us into an Alice-in-Wonderland world where probabilities can acquire negative values. Our judicious choice of variables, however, comes to the rescue; $x \to 0$, or $\zeta \to \infty$, corresponds to approaching this singularity but takes us no further; that is to say, all momenta are automatically cut off at a value determined by the position of the ghost pole ($\simeq 4\times10^{41}$ GeV). Now the loop expansion for β_1 will break down in the neighborhood of this pole. We may reasonably expect that perturbation theory will be good if t is such that $\alpha_1(t)(\equiv g_1^2/4\pi) \lesssim 0.1$; this corresponds to $p_{max} \lesssim 4\times10^{37}$ GeV -- large enough to permit us to assume that its finiteness has no effect on the physics we study -- and $t_{max} \lesssim 82$. Thus t must be deemed to be "large" if it is O(100).

When all is said and done, however, the fact remains that our considerations involve a cut-off; large though it is, it is still an undesirable feature in a description of electroweak interactions that we would like to describe as consistent. In this sense, the logic underlying the preceding development is flawed, and the theoretical scenario we have constructed must be regarded as provisional.

We have treated the electroweak system as an isolated system, the canonical theory as a fundamental theory. An alternate viewpoint would be to regard the Salam-Weinberg theory as an effective low energy theory, the fundamental theory being a grand unified theory or, better still, a super unified theory which embraces gravity as well. As emphasized in the Introduction (Section 1) such theories would also suffer from problems of consistency; indeed these problems are more complex than anything that afflicts electroweak theory. Pending their elucidation and resolution, one may impose upon the Salam-Weinberg theory a tentative constraint, a consistency requirement of sorts, that the theory be not required to handle momenta in excess of say the Planck mass. This leads to an upper bound on the Higgs mass which, for $m_t < 80$ GeV, is approximately $\sqrt{2}$ times larger than the one derived above. Indeed all alternate theoretical approaches to the problem of Higgs mass, that are known to us, lead to upper bounds larger than ours.[22]

3.7 Concluding Remark

The message to real physicists, our experimental colleagues, is that it may be worthwhile to look for the elusive Higgs particle at masses $\sim$ 10 GeV - 150 GeV. This is a very interesting mass range in that it is also the abode of pseudo-Goldstone bosons in the hypercolor scenario[23]. Should any semi-weakly coupled spin-0 particle be found in this range, it would be an important discovery. Investigation of the detailed properties of such a particle could lead to a clarification of the nature of the Higgs mechanism and illuminate the fundamental structure of electroweak theory. The experimental effort would in fact be worthwhile, even if the theories that predict relatively light spin-0 modes fail to survive.

4. MORE ABOUT THE SHORTCOMINGS OF THE CANONICAL FORMULATION. QUESTIONS OF ELEGANCE AND TASTE.

A built-in feature of conventional QFD is a commitment to the use of elementary spin-0 fields to establish the physically relevant gauge-hierarchy; this way of implementing the Higgs mechanism has been used in most schemes which combine QFD and QCD in the framework of a grand unified theory (GUT); it is what we have described as the "canonical methodology". That the procedure is marred by Landau ghosts and concomitant paradoxes has been noted earlier. Even if one could rescue systems of self-coupled elementary scalars from the trap of triviality -- in the absence of a mathematically rigorous demonstration that the S-matrix is unity, there is always room for hope -- there exist a variety of aesthetic shortcomings that render the methodology quite unattractive. The following observations should suffice to make the point.

i) Physics is deemed to stem from a hierarchy of nested gauge groups such as

$$G_n \supset G_{n-1} \supset \ldots \supset G_1 \tag{4.1}$$

a) One starts with an assortment of Higgs fields, (b) couples them in a manner which is both renormalizable and G_n invariant, (c) chooses parameters in the Higgs-potential in such a way that "spontaneous" symmetry-breaking occurs, producing massless bosons whose numbers and transformation properties match those of the broken generators, so that all can be devoured, and (d) requires that this choice of parameters yield a hierarchy of Goldberger-Treiman contants for the would-be Goldstone bosons:

$$f_n^2 > f_{n-1}^2 > \ldots > f_1^2 . \tag{4.2}$$

Here f_m is a generic symbol for the constants associated with those bosons which lead to the symmetry-breaking step: $G_m \to G_{m-1}$; thus the hierarchy in Eq. (4.2) leads to that in Eq. (4.1), with G_1 unbroken at the tree level. Indeed, one has an obligation to verify that this Rube Goldberg edifice does not collapse in the presence of radiative corrections or intrinsically non-perturbative effects; while the former can be calculated, the latter do not lend themselves to any systematic treatment.

ii) Wilson[1] has underlined an unsatisfactory feature of all renormalizable theories with elementary spin-0 fields in 4 dimensions: unlike all other divergences, which are logarithmic[2], the self energies of zero spin particles diverge quadratically. The relationship between bare and renormalized masses is

$$\mu^2 = \mu_o^2 - \lambda\Lambda^2 \tag{4.3}$$

where Λ is an ultra-violet cut-off and λ is the coefficient of the ϕ^4 term in the Lagrangian. If, following Salam[3], we assume that a natural cut-off will emerge at the Planck mass, $\Lambda \sim 10^{19}$ GeV, the bare mass, μ_o, has to be fine tuned to a precision of about one part in 10^{19} if μ is to be of the order of a few Gev.

iii) In the standard model, one begins with four real spin-0 fields; three of these are absorbed in the Higgs mechanism, leaving us with

one physical spin-0 field. This seems economical, until one realizes that 16 of the 19 independent parameters which characterize the model originate in the Higgs sector.
iv) The situation gets worse, if not acute, if one embeds the standard model in a grand unified theory.

Consider the currently fashionable SU(5) synthesis. The simplest choice of Higgs fields which will accomplish

$$SU(5) \rightarrow SU(3)_C \otimes SU(2)_L \otimes U(1)_Y \rightarrow SU(3)_C \otimes U(1)_Q \qquad (4.4)$$

is a (complex) 5 and a (real) 24. Since 15 of these fields must be absorbed in the Higgs mechanism, we are left with 19 real physical Higgs fields. The number of independent parameters is 22, actually greater than in the standard model; of these, 21 originate in the Higgs sector. Unfortunately, the theory with just a 5 + 24 of Higgses does not seem capable of describing physical reality*; to have independent masses for all the fermions, it is necessary to add a (complex) 45; this would give us a total of 109 physical Higgs fields with a concomitant escalation in the number of parameters.

The Achilles heel of the canonical realization of SU(5) is the gauge hierarchy problem[5]. There are two distinct mass scales, at $M_W \simeq 100$ GeV and at $M_x \simeq 10^{14}$ GeV; the question is: How is one to understand the ratio $(M_W^2/M_x^2) \simeq 10^{-24}$? To generate such disparate masses the parameters of the Higgs sector, including radiative corrections, must be fine tuned to a precision of one part in 10^{24}. It is evident that we are dealing with a feature of the model which is not natural, in any sense of the word.

Higher groups such as SO(10) or E_6 do not lead to any amelioration of these difficulties. With E_6, for example, one has to start with 650 Higgs fields[6].
(v) Most canonical formulations lead to no new physics in the range 10^2-10^{14} GeV, a fact first noted by Glashow in the SU(5), context . Hence the appellation for this energy interval: Glashow's desert. Such a desert seems highly implausible on the basis of probabilities.

5. DYNAMICAL HIGGS MECHANISM

5.1 Introduction

There have been many attempts to generate symmetry-breaking in a dynamical way, without benefit of elementary scalars[1-5]; we turn to them because they offer possible avenues of escape from the canonical methodology.

To put our discussion in the proper historical perspective, let

*Without a 45 plet of Higgs fields, one finds relationships such as $(m_d/m_s) = (m_e/m_\mu)$, where the quark masses are the current masses; experimentally $(m_d/m_s) \simeq 0.04$ whereas $(m_e/m_\mu) \simeq 0.005$. See, for example, Georgi and Jarlskog[4].

us note that the effect customarily associated with the name of Higgs had in fact been discovered in 1958 by Anderson[6] in his study of the theory of superconductivity; indeed the work of Anderson provides the first, albeit nonrelativistic, example of a dynamical Higgs mechanism. Relativistic examples of this phenomenon (i.e. the spontaneous breakdown of local gauge invariance without the help of elementary spin-0 fields) are afforded by two dimensional QED, solved by Schwinger[7] in 1962, and the U(1) models of Cornwall and Norton[8] and of Jackiw and Johnson.[9] Whether these examples illuminate the nature of dynamical symmetry breaking in the real world is open to question. The absence of a Goldstone theorem in two dimensions[10] renders suspect any general conclusions that one may be tempted to draw from two-dimensional models, the fact that symmetry-breaking in the U(1) models is an ultra-violet[11] effect makes it difficult to regard these models as paradigms relevant to the physical gauge hierarchy problem. At this time, the most promising approach to a dynamical alternative is based on the hope that QCD and QCD-like theories lead naturally to a (genuinely) spontaneous breakdown of chiral symmetries in flavor space; the resulting Goldstone bosons can then provide the longitudinal degrees of freedom that are needed to generate mass for flavor gauge fields. In this approach symmetry-breaking effects stem from the infra-red sector and one can meaningfully talk of short-distance or high-energy regimes in which a given symmetry may be deemed to have been restored[12]

5.2 Flavor Symmetries in QCD

Some proven results and some plausible conjectures pertaining to flavor symmetries in QCD, which play a crucial role in our discussion, may be summarized as follows.

Let us consider the Lagrangian:

$$L_{QCD} = -\frac{1}{8}\,\mathrm{Tr}G_{\mu\nu}G^{\mu\nu} + \sum_f \bar{\psi}_f i\gamma^\mu D_\mu \psi_f - \sum_f m_f \bar{\psi}_f \psi_f \qquad (5.1)$$

where G is the Yang-Mills curl of the suitably normalized color-gluon field matrix, D_μ is the gauge-covariant derivative and the summation is over all N_F quark flavors.

In the limit in which all $m_f \to 0$, the accidental global invariance group of L_{QCD} in flavor space is

$$G_{QCD}(\mathrm{ACC.}) = SU(N_F)_L \otimes SU(N_F)_R \otimes U(1)_A \otimes U(1)_V \qquad (5.2)$$

The QCD vacuum is presumed to be such that the vector and axial-vector charges associated with the first two groups in Eq. (5.2) satisfy

$$\dot{Q}_i^V = 0\ ,\quad Q_i^V|0\rangle = 0 \qquad (5.3)$$

$$\dot{Q}_i^A = 0 \quad , \quad Q_i^A|0\rangle \neq 0 \tag{5.4}$$

where $i = 1, 2, \ldots, N_F^2 - 1$. Failure of the conserved axial charges to annihilate the vacuum implies the existence of $N_F^2 - 1$ Goldstone bosons and a dynamical or Nambu-Goldstone mass, M, for the quarks. That the mechanism leading to Eq. (5.4) is intrinsically nonperturbative is most easily seen by noting that M satisfies a homogeneous renormalization group equation; this equation can be solved in the limit of small coupling[13]:

$$M \propto \mu \exp\left(\frac{-c}{g^2}\right) \tag{5.5}$$

where μ is the scale-parameter introduced to define the Green's functions of the theory, g is the renormalized QCD coupling constant and c is a known positive constant. It is possible, but has not been established to everyone's satisfaction, that Eq. (5.4) obtains in the presence of instantons[14]; at this time, it must be regarded as a plausible conjecture and its validity will be assumed in what follows. Note that f_π, the pion decay constant, also satisfies an homogeneous renormalization group equation; its g dependence is therefore identical to that of M.

The generator of the $U(1)_A$ factor in Eq. (5.2) also fails to annihilate the vacuum; however the symmetry is broken explicitly by the 't Hooft mechanism[15] and the would-be flavor-singlet Goldstone boson becomes massive -- a process which will hereinafter be referred to as 't Hooftization[16]. Finally we note that the $U(1)_V$ factor in Eq. (5.2) merely corresponds to conservation of baryon number by the strong interactions and is not relevant to our considerations.

Now let us turn on the mass terms in the Lagrangian. The mass m_f is the mass that occurs in current algebra calculations and which may, in accordance with customary usage, be described as "current mass". [In QCD, as in QED, $m_f\bar{\psi}_f\psi_f$ is a finite operator; mass renormalization is therefore multiplicative and trivial and we shall not explicitly distinguish between bare and renormalized masses.] The net quark mass M_f, the so-called constituent mass, is a superposition of the current and the dynamical masses:

$$M_f = m_f + M + \ldots \tag{5.6}$$

where the dots encompass terms such as the change in dynamical mass induced by the presence of current mass.

In any <u>specific gauge</u>, a precise definition of a momentum-dependent quark mass is furnished by

$$M_f(p) = \frac{1}{4} \mathrm{Tr}\, S'(p)_f^{-1} \tag{5.7}$$

where S' is the renormalized quark-propagator. The dynamical component of this mass is expected to damp rapidly (i.e. as a power of p) at high frequencies

$$M(p) \propto M \cdot (\frac{p^2}{\mu^2})^{-A} \qquad (5.8)$$

whereas the current component varies only logarithmically

$$m_f(p) \propto m_f \cdot \{\ell n\, (\frac{p^2}{\mu^2})\}^{-B} \qquad (5.9)$$

Here A is a dynamical parameter; B, however, is calculable in perturbation theory. Eqs. (5.8) and (5.9) permit us to refer to M and m_f as infra-red and ultra-violet masses respectively.

With the m_f's different from zero, the Goldstones --identifiable with the ordinary π's, K's etc. -- pick up mass. Clearly PCAC is a useful notion for flavors such that $m_f << M$. Since pion-PCAC is known to be good, we may assume that constituent u and d quark masses[17] provide us with a measure of the dynamical mass M:

$$M \simeq 3f_\pi \simeq 300 \text{ MeV} \qquad (5.10)$$

Eq. (5.10) gives us the natural mass scale of QCD.

While the rest of the discussion will be predicated on the premise that the structure envisaged above -- for flavor symmetries in QCD -- carries over mutatis mutandis to other QCD - like theories, it should be stressed that our formulation can in no sense be regarded as complete; a major shortcoming is that it does not take account of the phenomenon of color-confinement.

6. HYPERCOLOR THEORIES

6.1 The Hypercolor Scenario: An Artist's Conception

The simplest strategy for dynamical generation of Goldstone bosons -- which can fulfill the needs of QFD and, hopefully, resolve the full gauge-hierarchy problem -- is to introduce a new species of quarks which come in at least two flavors with vanishing current-masses and are characterized by at least two, possibly four or more, new[1] colors C'. Strong interactions among the new quarks are generated by gauging $G_{C'}$, the group associated with this (color)' degree of freedom; as indicated earlier, the structure of the resulting QC'D may be gleaned from analogy with QCD.

The (color)' degree of freedom, first introduced by Weinberg[2], has come to be known by a variety of names. Following Ref. 3, we shall call it hypercolor (HC); the terminology is convenient, with words such as hyperquark, hyperpion, hypersigma, etc. having an obvious meaning.

The strategy of hypercolor is based on the assumption that chiral symmetry in QCD-like theories with zero current quark-mass is realized in the Nambu-Goldstone way. A feature of this realization is the spontaneous generation of a mass scale which manifests itself via dynamical masses of quarks ($\sim$ 300 MeV for QCD) and non-vanishing values of the Goldberger-Treiman constants of Goldstone bosons such as the pion ($f_\pi \sim$ 100 MeV). Quite obviously the mass-scale of QCD is not adequate for endowing the weak bosons, $W^\pm$ and Z, with masses $\sim$ 100 GeV. One therefore postulates hyperfermions with the attribute of hypercolor; gauging of this hypercolor degree of freedom leads to QC'D, a QCD-like theory with a natural mass scale such that for hyperfermions $f_{\pi'} \lesssim$ 250 GeV. Such hyperpions, if they are true Goldstone bosons, can indeed play the role of unphysical Higgs fields, and thereby get the physical Higgs mechanism off the ground.

In the first step of the hypercolor scenario one starts with a gauge group $G = U(1) \otimes SU(2)_L \otimes SU(3)_C \otimes G_{C'}$, $G_{C'}$ being the unbroken hypercolor group. The hyperquarks, presumed to come in $N_{C'}$ colors and an <u>even</u> number of flavors, $N_{F'}$, are assigned weak isospins so that there are $N_{F'}/2$ left-handed doublets and $N_{F'}$ right-handed singlets.

The weak and electromagnetic interactions of hyperquarks may be taken to be isomorphic to those of ordinary quarks, so that a flavor doublet such as (u', d') transforms under the electroweak gauge group in the same way as (u, d) or (c, s), etc.

With a single massless flavor doublet of hyperquarks, and $SU(2)_L \otimes U(1)_Y$ as the electroweak group, the dynamical mechanism operates as follows. There are three massless hyperpions (the hyper-η is rendered massive by the 't Hooft mechanism); these mix with $W^\pm$ and Z which therefore acquire mass. The isotopic and hypercharge properties of the hyperpions, in conjunction with the isotopic-spin invariance of QC'D, imply:

$$M_Z^2 \cos^2\Theta_W = M_W^2 = \left(\frac{e}{2\sin\Theta_W}\right)^2 f_{\pi'}^2, \tag{6.1}$$

where Θ_W is the Glashow-Weinberg-Salam angle ($\sin^2\Theta_W \simeq 0.22$) and $f_{\pi'}$ is the hyper-pion decay constant.

Consider next the case of several hyperquark doublets. While only the doublet (or doublets) with zero current mass can trigger the Higgs mechanism, i.e. move gauge-field masses from zero to finite values, doublets with nonzero current masses can enhance the masses of those gauge fields which have already acquired a finite mass. If we assume that all hyperpions -- irrespective of whether they belong to the same or different isotopic multiplets -- have equal decay constants and that they have masses which are either zero or negligible compared to W and Z masses, we may extend Eq. (6.1) as

$$M_Z \cos\Theta_W = M_W = \frac{e}{2\sin\Theta_W} f_{\pi'} \sqrt{N_{F'}/2} \tag{6.2}$$

Note that our assignment of hyperquarks to representations of the electroweak group is sufficient to ensure that the well-verified relationship between W and Z masses, a consequence of the use of Higgs <u>doublets</u> in the Weinberg-Salam model and often called the

$\Delta I_{weak} = \frac{1}{2}$ rule, emerges naturally. Indeed, in so far as W and Z masses are concerned, the hypercolor scheme is equivalent to using an "elementary" Higgs doublet in the effective Lagrangian[2,3]

$$L_{eff} = (D_\mu \phi)^\dagger (D^\mu \phi) + \ldots, \qquad (6.3)$$

where D_μ is the appropriate gauge-covariant derivative,

$$\phi = \begin{pmatrix} i\pi'^{+} \\ \frac{\sigma' - i\pi'^{o}}{\sqrt{2}} \end{pmatrix} \qquad (6.4)$$

and

$$\sigma' = \langle\sigma'\rangle + \Sigma \qquad (6.5)$$

where

$$\langle\sigma'\rangle \equiv f_{\pi'} \sqrt{N_{F'}/2} \qquad (6.6)$$

and Σ is of mass [comparable to that of the 't Hooftized η'] $\sim 2\,\Lambda_{HC} \sim 600$ GeV. But it will be too broad to be visible as a particle.

Eqs. (6.3), (6.4) and (6.6) provide a concrete realization of the proposal that the Higgs fields of the canonical methodology be viewed as phenomenological props[4]. However, as we shall see later, the experimental signatures of elementary Higgses and the composite fields of the hypercolor scenario are not quite the same[3].

The hyperpion decay constant, which determines the mass scale in QC'D, can be evaluated from Eq. (6.2):

$$f_{\pi'} = \frac{250}{\sqrt{N_F'/2}}\ \text{GeV} \qquad (6.7)$$

6.2 Problems and Tentative Solutions

Hypercolor-based DSB has been afflicted from the very beginning with rather serious problems. In this section we shall discuss the main problems and possible solutions to some of them.

A) In the absence of explicit Higgs fields, the accidental global invariance group of the Lagrangian is much larger than $G = U(1) \otimes SU(2)_L \otimes SU(3)_C \otimes G_{C'}$. For example, with four quark flavors the Lagrangian is invariant under additional "horizontal" groups such as

$$G_H = SU(2)_L \otimes SU(2)_R \otimes \ldots \qquad (6.8)$$

The generators of the groups indicated in Eq. (6.8) are:

$$Q_+ = \int d^3x\ (u^\dagger c + d^\dagger s)_{L\text{ or }R},\quad Q_- = Q_+^\dagger$$

$$Q_3 = \frac{1}{2}[\,Q_+,\ Q_-\,] \tag{6.9}$$

One may verify without difficulty that these groups commute with the electroweak group. The full horizontal group is specified more precisely in Eq. (6.17), below.

Some of these extra flavor symmetries are broken, but only spontaneously, and therefore lead to unwanted Goldstone bosons.

B) The second serious problem facing the scheme stems from the failure of the ϕ's to couple directly to fermions without the attribute of hypercolor. There is thus no mechanism for generation of lepton mass or of current quark mass.

C) There is the problem of calculational intractability, common to all theories in which symmetry-breaking is dynamical; the use of phenomenological Higgs fields ϕ is a start towards a resolution of this problem.

Attempts have been made to resolve the problems which afflict the hypercolor-based dynamical scheme; none can be said to have been terribly successful. To resolve (A), one may get rid of these bosons, by gauging these horizontal symmetries, and absorbing them in the Higgs mechanism. The relevant gauge fields must, however, become so massive that these horizontal interactions do not disturb the phenomenology of weak interactions -- reproduced so well by the Weinberg-Salam model.

A satisfactory solution to problem (B) seems even more elusive. The simplest way to generate current quark masses is to enlarge the gauge group in such a way as to have currents which mix "color" and "hypercolor"; this first attempt at resolution, due to Dimopoulos and Susskind[6], introduced and utilized the notion of "Extended Hypercolor" (EHC) group.

It is supposed that the EHC group has generators associated with transitions from ordinary fermions to hyperfermions. The gauge bosons associated with these generators are required to generate light fermion masses. To see how this happens, consider the graph of Fig. 6.1a.

Quark masses stemming from this mechanism may be crudely estimated -- for a more refined treatment, vidê infra -- as

$$m_q \sim \frac{g^2}{8\pi^2}\ \frac{\langle 0|\bar{q}'q'|0\rangle}{M^2} \tag{6.10}$$

where the hyperfermion condensate, $\langle 0|\bar{q}'q'|0\rangle$, may be related to the quark condensate, $\langle 0|\bar{q}q|0\rangle$, by assuming that QC'D is a scaled-up version of QCD -- as postulated in Ref. 3.

Unfortunately, the EHC-construct generates new problems; two of these have been widely discussed in the literature.

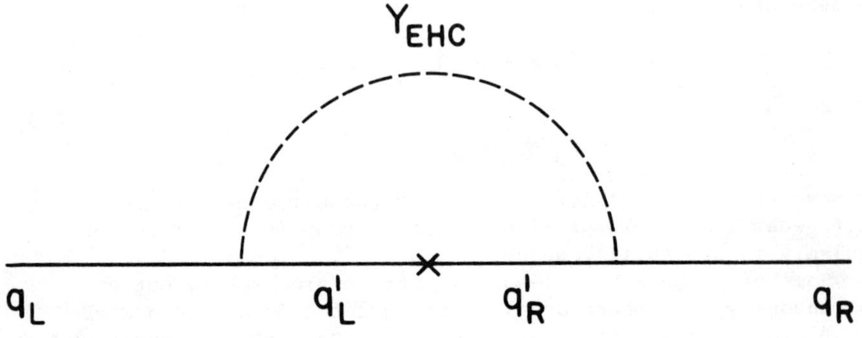

Fig. 6.1a. The mechanism whereby EHC bosons give masses to conventional fermions.

D) First and foremost is the menace of flavor-changing neutral currents (FCNC's), and consequent uncomfortably large $\Delta S = 2$ amplitudes. To see how this develops, note that the EHC group must contain the generators:

$$Q(s) = \int d^3x\, s_w^\dagger q' \qquad (6.11)$$

$$Q(d) = \int d^3x\, d_w^\dagger q' \qquad (6.12)$$

if the $s_w - d_w$ mass matrix is to be non-trivial, the w-suffix indicating that we are dealing with weak, as opposed to mass, eigenfields. Closure of the Lie algebra then requires that it also contain the generators:

$$Q_1 = [\, Q(s),\, Q(d)^\dagger \,] = \int d^3x\, s_w^\dagger d_w \,, \qquad (6.13)$$

$$Q_2 = Q_1^\dagger = \int d^3x\, d_w^\dagger s_w \,, \qquad (6.14)$$

$$Q_3 = [\, 1/2 \,] \int d^3x\, (s_w^\dagger s_w - d_w^\dagger d_w) \,. \qquad (6.15)$$

These generators imply FCNC's and $\Delta S = 2$; indeed, they account for the worst of the $\Delta S = 2$ problem through the single gauge boson exchange process, dubbed SGEX in Ref. 12. Making many assumptions, which need not be reviewed here, the authors of Ref. 12 found:

$$L^{\Delta S=2}_{eff.} \text{ (Hypercolor)}$$

$$\approx 10^3 \, L^{\Delta S=2}_{eff.} \text{ (Canonical)} + \text{(RR and LR)} \qquad (6.16)$$

for

$$m_s \approx 200 \text{ MeV}.$$

Here (RR and LR) indicates contributions to $L_{eff.}$ which are absent in the canonical theory; they stem from the coupling of right-handed currents to themselves and to left-handed currents; as is evident, both currents are required for mass generation in the dynamical construct.

The source of the difficulty lies in Eq. (6.10); to obtain adequate mass for the s-quark it is necessary to lower M; this enhances FCNC effects to an unacceptably high level.

E) Second is the possibility that the $\Delta I_{Wk} = \frac{1}{2}$ Rule, as expressed through the relationship $M_Z^2 \cos^2\Theta_W = M_W^2$, may be jeopardized. A crucial input in deriving this formula is the SU(2)-based result that $f_{\pi^+} = \sqrt{2}\, f_{\pi^0}$; if the EHC mechanism yields large mass splittings in quark doublets, it could also disrupt iso-spin relationships to the point where the W and Z masses no longer conform to $\Delta I_{Wk} = 1/2$.

Some of the above problems have been recently "solved". To appreciate the quote marks, let us consider (D) and (E) in the context of a precise formulation of the EHC construct for quark-mass generation.

6.3 Quark Current Masses:

Our discussion of quark masses utilizes the mechanism proposed for resolution of (A) to generate EHC and thence graphs of the type depicted in Fig. 6.1a. We start with the observation that, in the absence of Higgs couplings, the quark -- gauge-field sector is actually invariant under the group

$$G_H^{QUARK}: \; SU(n)_L^{d,u} \otimes SU(n)_R^d \otimes SU(n)_R^u \otimes U(1) \qquad (6.17)$$

where the left-handed group operates in the space of both "up" and "down" quarks, the first right handed group operates only in the space of "down" quarks etc., n being the number of "up" or "down" quarks. To avoid unwanted Goldstone bosons G_H is gauged; quark masses which simulate current masses at low energies emerge as a bonus. Note that we have taken advantage of the fact that the right handed group in Eq. (6.8) can be factored into two separate groups; this is to split the "up"-"down" degeneracy.

The quintessential features of the problem at hand are best brought out by making some simplifying assumptions. Let us work with just four quark flavors and assume that hypercolor - bearing quarks, the hyperquarks, are devoid of the attribute of ordinary color. The following assignments under the horizontal group then suggest themselves: $(d_W, s_W, d'_i)^T_L$ and $(u_W, c_W, u'_i)^T_L$ transform as n-plets under $SU(n)^{d,u}_L$, $(d_W, s_W, d'_i)^T_R$ and $(u_W, c_W, u'_i)^T_R$ under $SU(n)^d_R$ and $SU(n)^u_R$ respectively. Here i (= 1,2... $n'_C n'_F$) labels the hyperquarks, $n \equiv 6 + n'_C n'_F$, n'_C and n'_F being the number of hypercolors and hyperflavors; the suffix W, omitted on the hyperquarks, indicates that the fields are weak, rather than mass, eigenfields. The resulting fermion -- gauge-field interaction may be written as:

$$\begin{aligned}
\delta L_I = g_1 \, [& (\bar{d}'_i \gamma_\mu d_W + \bar{u}'_i \gamma_\mu u_W)_L Y_1^{i\mu} \\
& + (\bar{d}'_i \gamma_\mu s_W + \bar{u}'_i \gamma_\mu c_W)_L \cdot Y_2^{i\mu}] \\
+ g_2 \, [& (\bar{d}'_i \gamma_\mu d_W)_R \cdot Y_3^{i\mu} + (\bar{d}'_i \gamma_\mu s_W)_R \cdot Y_4^{i\mu}] \\
+ g_3 \, [& (\bar{u}'_i \gamma_\mu u_W)_R \cdot Y_5^{i\mu} + (\bar{u}'_i \gamma_\mu c_W)_R \cdot Y_6^{i\mu}] \qquad (6.18) \\
& + \text{terms diagonal in quark and hyperquark spaces + H.C.}
\end{aligned}$$

where the g's are the coupling constants for the three SU(n) groups. With the additional assumption of no hyperflavor mixing, the gauge fields with well defined mass are related to the fields in Eq. (6.18) via

$$\bar{Y}_\ell^{i\mu} = a_{\ell m} Y_m^{i\mu} \qquad (6.19)$$

a being a transformation of the group SO(6):

$$a_{\ell m} a_{\ell n} = a_{m\ell} a_{n\ell} = \delta_{mn} \qquad (6.20)$$

The current mass of, say, the s-quark can now be calculated from Fig. 6.1b; using the procedure described in Ref. 18, for handling the momentum dependence of the dynamical hyperquark mass, one obtains

$$\begin{aligned}
m_s = \Sigma_\ell \, a_{\ell 2} \, a_{\ell 4} \, & (\delta M_\ell^2 / M^2) \\
& (n'_F n'_C g_1 g_2 / 4\pi^2) \cdot m_D^3 / M^2 \\
& \ln (M^2 / m_D^2) \qquad (6.21)
\end{aligned}$$

where $\delta M_\ell^2 \equiv M_\ell^2 - M^2$, M being the mean gauge-field mass, and $m_D \equiv m_D(\Lambda^2_{QC'D})$. Eq. (6.21) reduces to

$$m_s \simeq 12 \text{ GeV}(\Sigma_\ell \, a_{\ell 2} \, a_{\ell 4} \, \delta M_\ell^2/M^2) \tag{6.22}$$

for rather reasonable values of the various parameters: $m_D \simeq 1$ TeV, $M \simeq 10$ TeV, $n'_F n'_C \simeq 10$; $g_1 \simeq g_2 \simeq 1$.

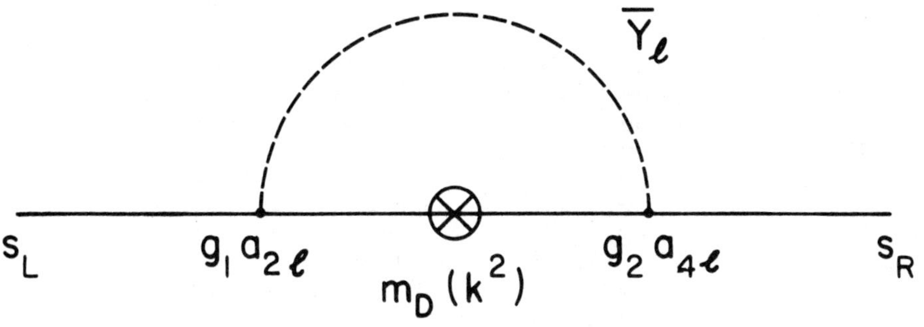

Fig. 6.1b. Graph contributing to the mass of the s-quark.

The main point, brought out by Eq. (6.22), may now be stated as follows: To get adequate masses for the heavier quarks, it is not necessary to lower M to values that may be unacceptable to weak interaction phenomenology, contrary to the fears of some early explorers.[12] Most widely cited was the possibility of dangerously large couplings involving FCNCs and consequent $\Delta S = 2$ amplitudes of an unacceptable magnitude; to a reexamination of these topics let us now turn.

6.4 Flavor Changing Neutral Currents

Consider a quark-space diagonal piece of the Lagrangian

$$\delta L = g_2 \, [\, \overline{D}_w(\Theta_1)\gamma_\mu \, \tfrac{1}{2} \, \tau_i D_w(\Theta_1)]_R \cdot E_R^{\mu i} \tag{6.23}$$

where $D_w(\Theta_1)^T = (d_w, s_w)$. This interaction arises from the SU(6) sub-group of the $SU(n)_R^d$ factor in the gauge group; the piece displayed corresponds to retaining only the color singlet field in the reduction of SU(6) gauge fields under $SU(3)_C \otimes SU(2)_F$: 35 = (1,3) +

(8,1) + (8,3). Now if $E^{\mu 1}$ and $E^{\mu 3}$ mix, and if the mixing angle be 2χ, the quark couplings to mass-eigenfields are $D_W(\Theta_1-\chi)$. The effective $\Delta S = 2$ interaction stemming from (6.23) through the exchange of one gauge boson is thus

$$\delta L^{\Delta S=2}_{\text{eff.}} = (g_2^2/4M^2)(\delta M^2/M^2) \cdot \sin^2 2(\Theta_1-\chi)$$

$$\times (\bar{d}\gamma_\mu s)_R \cdot (\bar{d}\gamma^\mu s)_R + \text{H.C.} \tag{6.24}$$

where $\delta M^2 = M(\bar{E}^{\mp})^2 - M(\bar{E}^3)^2$. Note that the Cabibbo angle is $\Theta = \Theta_1-\Theta_2$, where Θ_2 is the mixing angle in the (u,c) sector; the formula analogous to Eq. (6.24) for $\Delta C = 2$ would involve Θ_2.

The strength of FCNC couplings is thus determined by angles such as $\Theta_1-\chi$ which can not be inferred from low energy physics; they may be calculable, in principle, but present techniques are not adequate for the task; in the present state of the art, one can regard them as adjustable free parameters.

More interesting than this interim solution of the $\Delta S = 2$ problem is the natural suppression of such amplitudes through what has been called[11] Pseudo-GIM. This effect has its origins in the $SU(2)_F$ symmetry encountered earlier; in the exact symmetry limit $m_s = m_d$, $m_c = m_u$, $\delta M^2 = 0$ and the $\Delta S = 2$ amplitude vanishes; the factor $\delta M^2/M^2$ in Eq. (6.24) thus corresponds to pseudo-GIM suppression. The mechanism has been shown to persist up to the one - QFD - loop level[18]; it thus helps keep the double gauge boson exchange contribution to $\Delta S = 2$ at a level no worse than in the canonical theory.

6.5 The $\Delta I_{wk} = 1/2$ Rule

In the canonical theory, the Higgs sector has an O(4) - symmetry ($\equiv SU(2)_L \otimes SU(2)_R$) which collapses to $SU(2)_{L+R}$ after spontaneous symmetry breaking; this custodial SU(2) guarantees the $\Delta I_{wk} = 1/2$ rule,

$$\rho \equiv (m_W^2/m_Z^2 \cos^2\Theta_W) = 1 \tag{6.25}$$

to all orders in the scalar coupling. Furthermore, $\rho = 1$ and $m_u \neq m_d$ are compatible with each other at the tree level. At the one-loop level one obtains the $\Delta\rho$ of Eq. (2.19).

The situation in the hypercolor scenario is very different.[23] Quark mass generating interactions break custodial SU(2) and imply SU(2) - breaking interactions among the hyperquarks. Corrections to the $\Delta I_{wk} = 1/2$ rule may thus be linear in the mass splittings; indeed a crude estimate, based on evaluation of graphs such as the one shown in Fig. 6.2, yields[23]

$$\Delta\rho \simeq (m_u - m_d)/8\pi^2\Lambda_{QC'D} \tag{6.26}$$

While Eq. (6.26) does not give any immediate cause for alarm, better measurements and calculations* of $\Delta\rho$ could lead to a decisive confrontation with experiment. This expression of hope assumes, of course, that a reliable calculational procedure will be available in the not-too-distant future.

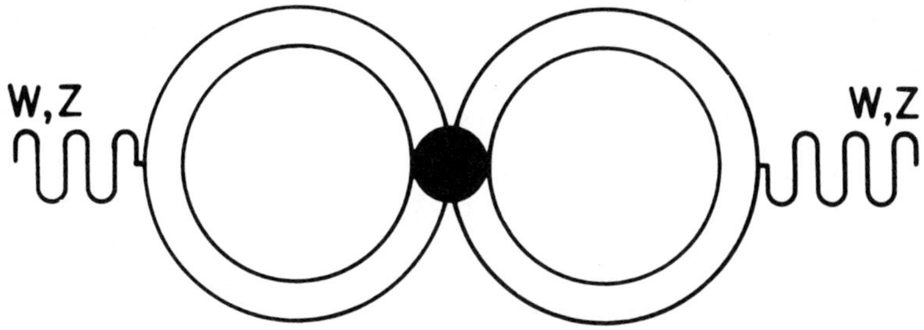

Fig. 6.2. Gauge boson self-energy diagrams which contribute to $\Delta\rho$ in hypercolor theories. The double lines represent hyperquarks.

6.6 Phenomenological Implications of Hypercolor

The DSB theories under discussion are easily distinguishable[25] from the normal Higgs field scenario at energies greater than one to two TeV. If nature makes use of hypercolor, a new hadronic spectroscopy -- rich in content and carrying the signature of this brand of DSB -- will become manifest. The challenge is to pinpoint the nature of the Higgs mechanism via experiments in the 10 GeV to 100 GeV range, i.e., at energies which are either accessible or likely to be in the near future.[3,25,26]

To explore the spectroscopy of hyperhadrons, i.e. hypercolor-singlet hyperquark-antihyperquark or multi-hyperquark states, it is necessary to know - or, at least, have a lower bound on- the number of hypercolors and have an educated guess for constituent hypercolor masses.

The number of hypercolors may be constrained by requiring that there exist a finite grand-unification mass M at which the QC'D and QCD coupling constants are of the same order:

$$\bar{g}_{C'}(M) \approx \bar{g}_{C}(M) \tag{6.27}$$

*To see another source of contributions to $\Delta\rho$ within hypercolor theory see also A. Zepeda[24] and A. Hernández and A. Zepeda (to be published).

If $M > \Lambda_C$ and Λ'_C, the characteristic mass scales in QCD and QC'D respectively, Eq. (6.27) is equivalent to

$$(11N'_C - 2N'_F)\ln(M/\Lambda'_C) \approx (11N_C - 2N_F)\ln(M/\Lambda_C) \qquad (6.28)$$

where N'_C and N'_F are, respectively, the number of colors and flavors in QC'D - tacitly assumed to be an $SU(N'_C)$ gauge theory. In QCD, $N_C = 3$ and we shall assume that $N_F = 6$. Since $\Lambda'_C >> \Lambda_C$, it is evident that Eq. (6.28) can be satisfied only if

$$11N'_C - 2N'_F > 21. \qquad (6.29)$$

Thus $N'_C \geq 3$ if $N'_F \leq 4$; if $N'_F \geq N_F$, $N'_C \geq 4$. Needless to add, sharper constraints would be obtained if one were to specify the grand-unification group G and the nature of $SU(N'_C)$ embedding in G. In the following, where a specific value is needed, we shall take $N'_C = 4$.

To determine Λ'_C, which sets the scale of constituent hyperquark masses, one may postulate that QC'D is a scaled-up version of QCD, and equate the dimensionless ratios

$$(f_{\pi'}/\Lambda'_C\sqrt{N'_C}) = (f_\pi/\Lambda_C\sqrt{N_C}) \ . \qquad (6.30)$$

Here $\sqrt{N_C}$ and $\sqrt{N'_C}$ are statistical color factors. [The simplest argument for the occurrence of these factors, in the manner in which they do in Eq. (6.30) proceeds as follows. The charge-raising axial current associated with a prescribed flavor-doublet (u, d) is $A_\mu^{(-)} = \Sigma_\alpha \bar{u}_\alpha \gamma_5 \gamma_\mu d_\alpha$ the summation being over all colors; the state vector of the pion with the same flavor quantum numbers is $(1/\sqrt{N_C})\Sigma_\alpha |u_\alpha \bar{d}_\alpha>$. Hence, in a parton-like limit, $f_{\pi^+} \equiv f_\pi\sqrt{2} = fN_C/\sqrt{N_C}$, where f is a "reduced" Goldberger-Treiman constant for a given flavor and color. Our scaling hypothesis is the statement that $(f'/\Lambda'_C) = (f/\Lambda_C)$.] Eqs. (5.11), (6.7) and (6.30) imply

$$\Lambda'_C \approx 750(6/N'_C N'_F)^{1/2}\text{GeV} < 750 \text{ GeV} \qquad (6.31)$$

the inequality following from the model-independent constraint on QC'D: $N'_C N'_F \geq 6$. To assess orders of magnitude, without getting involved in specific models, it is not unreasonable to set $\Lambda'_C \sim 0.5$ TeV.

Several facets of hyperhadron spectroscopy can now be discussed.

(i) Hyperbaryons:

These are, by definition, multi-hyperquark states transforming according to the completely antisymmetric singlet representation of $SU(N'_C)$. The expected masses are, therefore, in the neighborhood of

$$M_B \sim N'_C \Lambda'_C \approx 1.8(N'_C/N'_F)^{1/2}\text{TeV}. \qquad (6.32)$$

Conservation of hyperbaryon number is expected to be on the same footing as conservation of ordinary baryon number; consequently, hyperbaryons have to be pair-produced. Experimental searches for these objects must therefore await the construction of machines yield ing $\sim$ 4 TeV in the center-of-mass frame. In the meantime, hyperbaryons offer a fertile field for theoretical conjecture and speculation, and some interesting problems for cosmologists. Perhaps the most important question stems from the possibility that hypernucleons (defined to be the lowest mass hyperbaryons) and hypernuclei may constitute new islands-of-stability in the hadronic spectrum in the trans-TeV region. A substantial amount of such matter should have been created in the early stages of evolution and one may legitimately ask: where is the hypermatter?

(ii) Hypervector mesons:

If the current masses of quarks and hyperquarks are of the same order of magnitude - a feature found in many proposed models for mass-generation - almost all hyperhadronic mass is dynamical. This means that the wide disparity between ρ - ω, J/ψ and Υ masses is unlikely to be duplicated in the hypermeson spectrum; one expects, rather, that all vector meson masses would be close to

$$M_{(VM)'} \approx 2\Lambda'_C . \qquad (6.33)$$

The hypervector mesons will undoubtedly decay fairly rapidly into relatively light pseudoscalars (see below); however, a credible estimate of the widths is not available at this time. An e^+e^- machine with 0.5 - 1 TeV beams (super LEP?, next phase of SLC?) would afford the best means for producing and studying these 1^- hypermesons.

(iii) Hyper-scalar mesons:

The set of 0^+ hypermesons (mass $\sim$1 TeV) includes the σ', the precise analogue of the left-over Higgs of the standard model. However - unlike the Higgs particle - the σ' will decay very rapidly into light pseudoscalars and, in consequence, will be much too broad to be visible as a distinct particle.

(iv) Hyper-pseudoscalar mesons and the pseudo-Goldstone sector:

Pseudoscalar mesons are, by far, the most important particles in the hyperhadron spectrum; their importance stems from the fact that they are the lightest hyperparticles - the only ones that can be produced at energies which, if not currently available, are likely to be available in the near future.

In the absence of any mechanism for generating current hyperquark masses which are anywhere more than a small fraction of the dynamical masses, the pseudoscalar mesons are best viewed in terms of the Nambu-Goldstone phenomenon-rather than as $^1S_0$ states of a naive hyperquark model. The modes generated by the QC'D Lagrangian, in the presence of electroweak and other (sub-electroweak) interactions, may be categorized as follows: (a) True Goldstone bosons; these - unless they are electrically neutral and have other special properties such that they can not be ruled out on the basis of existing experiments - should not occur. (b) True (would be) Goldstone bosons, absorbed in the Higgs mechanism. (c) Goldstone bosons subject to 't Hooftization. (d) Pseudo-Goldstone[3,27] bosons (PGB's) which acquire mass through the agency of electroweak interactions. It is this last category which concerns us here.

The masses of the PGB's have been estimated by several authors[3,28,12,30]; various uncertainties and / or model dependent features are to be found in all calculations; however, all published values lie in the range 10 GeV-100 GeV.

The electromagnetic radius of charged pion-like PGB's, the hyperpions $\pi'^{\pm}$, may be inferred from hypervector dominance arguments; it is $\sim\sqrt{6}(2\Lambda')^{-1} \sim (0.4\ \text{TeV})^{-1}$. Consequently at energies of a few tens of GeV, the hyperpions are point-like and should be produced in e^+e^- annihilation; their contribution to the usual R-parameter is[3]:

$$\Delta R = \frac{1}{4} \sum_{\pi'} Q_{\pi'}^2 \,(1 - 4m_{\pi'}^2/s)^{3/2} \tag{6.34}$$

where the summation extends over all hyperpions with charges $Q_{\pi'}$ which can be produced at c.m. energy $\sqrt{s}$. Methods for identifying $\pi'^{\pm}$ have been discussed elsewhere[29]; notable for its complete and thorough treatment is the work of Ali, Newman and Zhu[31].

If charged hyperpions are seen at a low mass value, say 10 GeV, it would be a strong indication that one is not on the wrong track. However, one can not really tell a π'^+ from ϕ^+, one of the charged Higgs particles which will be found in any extension of the standard model. To obtain observational insight into the nature of the Higgs mechanism, it is necessary to investigate the neutrals. For distinguishing between π'^o, a light neutral PGB of the hyper-color scenario, and ϕ^o, the left-over Higgs scalar of the canonical theory, two handles are available[32]:

(a) In processes which involve only flavor-diagonal couplings of spin-0 particles to fermions there are the parity tests of Ref. 3. Here it is worth pointing out that the considerations of Ref. 3 are model-independent if π'^o is a flavor-neutral PGB. Such PGB's transform as ${}^1S_o$ bound states of hyperquarks q_i' and anti-hyperquarks $\overline{q}_i'$ and are therefore odd under CP. Thus CP invariance, which may be deemed to be good for our purposes, is sufficient to guarantee: $\phi^o \to D\overline{D}$ (yes), $\pi'^o \to D\overline{D}$ (no), etc.

(b) There are no PGB-gauge-field couplings at the no-QFD-loop level[3]. This is to be contrasted with the situation in standard QFD which predicts a large ϕZZ coupling at the tree level.

To take advantage of (b) and (a) above, in the most efficient way possible, it is necessary to consider experiments at high-luminosity colliders. The reactions which warrant careful scrutiny, at LEP, SLC, TEVATRON and SSC for example, would be as follows:

LEP

$$e^+e^- \to J_T \to \pi'^o + \gamma \tag{6.35a}$$

$$e^+e^- \to J_T \to \phi^o + \gamma \tag{6.35b}$$

$$e^+e^- \to Z \to \pi'^o + \mu^+\mu^- \tag{6.36a}$$

$$e^+e^- \to Z \to \phi^o + \mu^+\mu^- \tag{6.36b}$$

$$e^+e^- \to "Z" \text{ or } "\gamma" \to \pi'^o + Z \tag{6.37a}$$

$$e^+e^- \to "Z" \to \phi^o + Z \tag{6.37b}$$

TEVATRON AND SSC experiments[30,34]

$$pp \to J_T + X$$

$$\hookrightarrow \begin{cases} \pi'^o + \gamma & (6.38a) \\ \phi^o + \gamma & (6.38b) \end{cases}$$

$$pp \to \pi'^o + Z + X \tag{6.39a}$$

$$pp \to \phi^o + Z + X \tag{6.39b}$$

$$pp \to \pi'^o + X \tag{6.40a}$$

$$pp \to \phi^o + X \tag{6.40b}$$

Here "Z" means a virtual Z. The dominant mechanism for the last reaction is two-gluon fusion; the same mechanism dominates (6.40a) in models in which hyperquarks carry the attribute of color.

Cross sections for the reactions listed above have been calculated by various authors[30,32,34]. For any reasonable choice of parameters, ϕ^o and π'^o are produced at comparable rates in reactions (6.35), (6.38) and (6.40); the CP-tests (a) would therefore be useful for distinguishing between them; furthermore, the cross sections for (6.36a), (6.37a) and (6.39a) are expected to be negligible compared to those for (6.36b), (6.37b) and (6.39b) respectively. With an unreasonable choice, one can enhance the cross section for π'^o production to the point where (6.36a) and (6.36b), for example, proceed at comparable rates; the dimuon invariant mass distribution, however, is very different in the two cases[32].

The ratio $\Gamma(Z \to \pi'^o + \mu^+\mu^-)/\Gamma(Z \to \phi^o + \mu^+\mu^-)$ is plotted in Fig. 6.3 for equal values of π'^o and ϕ^o mass. It is evident that unless one is willing to consider absurdly large values of $N_{c'}$ and $N_{F'}$, a positive signal in $Z \to \mu^+\mu^- +$ (spin 0) would be indicative of ϕ^o. Even if one is willing to consider values of $N_{c'}$ and $N_{F'}$ such that the signals for π'^o and ϕ^o production are of comparable magnitude, one can tell which particle is being produced by comparing the dimuon invariant mass distributions. (See Fig. 6.4).

We conclude by stating some of the morals which can be drawn from our discussion of hyperhadron spectroscopy.

First, PEP and PETRA, which have the energy to produce relatively light charged hyperpions, can tell us if - in taking hypercolor seriously - we are on the right track.

Second, LEP and TEVATRON could provide information about spin-0 neutrals that would help unravel the nature of the Higgs mechanism.

Third, nature could choose to be perverse and make use of spin-0 particles whose masses are such as to give us no relevant positive signal either at PEP or at PETRA or at LEP or TEVATRON. Experiment would then yield no information about the mechanism of symmetry breaking until multi-TeV machines are built.

6.7 Present Status of Hypercolor

The preceding discussion permits us to appraise the present status of the hypercolor-based dynamical alternative to the canonical theory.

(a) The elemental problems that afflicted the scenario in its early days are soluble, but--apart from an occasional glimmer of elegance, such as provided by pseudo-GIM--the solutions can not exactly be described as attractive.

(b) Further progress is impeded by the mathematical intractability of the formalism, a shortcoming common to all schemes for dynamical symmetry breaking. While almost everything is calculable, in principle, nothing can be calculated with present technology. Parameter proliferation can not be checked.

(c) As indicated earlier, hypercolor can play a useful role only in the context of grand unified theories based on semi-simple groups. To date, however, it has not been possible to construct a satisfactory working model--one in which the symmetry-breaking pattern may be deemed to be physically acceptable.

(d) Perhaps the only bright feature of hypercolor is its immediate experimental relevance. Despite the apparent insolubility of the underlying theoretical problems, it is possible to extract some well defined predictions; these should certainly be tested. One can then either nail the coffin -- without some experimental support it would be hard to take the subject seriously -- or learn to live with the construct and devise ways and means of molding it into an acceptable theory.

From composite models of Higgs bosons to those of quarks and leptons seems a natural step, and many authors have taken the plunge--some, indeed, with the hope that the difficulties of hypercolor may find amelioration in an all-composite framework. To this development we now turn.

7. COMPOSITE MODELS OF QUARKS AND LEPTONS

7.1 Motivation

To motivate composite models of quarks and leptons[1], let us take stock of all the elementary particles in the canonical implementation of the SU(5) synthesis, and compare what we have to what we had in the watershed year 1964 - before the quark model came into vogue.

From Table 7.1 it is clear that despite tremendous strides in our understanding of elementary particle interactions in terms of the gauge theoretic method, the quest for an economical description of the structure of matter has been less than successful.

7.2 The Quark Model Paradigm

To set the stage for a discussion of composite models, let us recall a rather successful model of nucleon structure, the so-called naive - but actually not-so-naive quark model. Historically, Bég, Lee and Pais were the first to discuss this model[2]; their purpose was to use the nuclear-physics experience to understand a result which they had derived in the framework of the SU(6) theory of Gürsey and Radicati[3], a relationship between the magnetic moments of protons and neutrons:

$$\mu_p/\mu_n = -\frac{3}{2} \tag{7.1}$$

Table 7.1. Spectrum of "elementary" particles in 1985 contrasted to that in 1964. Entries followed by a single question mark correspond to hitherto unobserved particles which are required by the canonical theory; entries followed by a double question mark correspond to the additional particles which would occur in SU(5)-based grand unification. The supergravity enthusiast may wish to delete the null entry in the 1985 column and insert his favorite numbers, followed by an appropriate number of question marks.

"Elementary" particles	1985	Before the quark model 1964
Spin $\frac{1}{2}$ fermions:		
Leptons	6	4
Hadrons	18	8
Spin 1 bosons:		
Massless	9	1
Massive	3 +12(??)	9
Spin 0 bosons	1(?)+108(??)	9
Spin $\frac{3}{2}$ fermions	0	10

Let us review the argument leading to Eq. (7.1) in modern language.

QCD confines quarks in a region of extension $\sim 0.7 \times 10^{-13}$ cms $\sim(300\ \text{MeV})^{-1}$. Now global chiral SU(2) ⊗ SU(2) symmetry emerges as an invariance of the QCD-Lagrangian in the limit in which the mechanical or current masses of the u and d quarks are zero. The absence of mass terms in the Lagrangian does not, of course, mean that the quarks are massless; since the chiral group is realized in the Nambu-Goldstone way[4], they have a dynamical or constituent mass. (The concomitant Goldstone bosons are the pions.) Indeed it is tempting to conjecture that this dynamical mass and the effective mass by virtue of confinement are one and the same, that confinement and spontaneous breakage of chiral symmetry are cognate phenomena. Nucleons are to be viewed as three-quark nuclei, loosely bound in the sense that $M_N \cong 3M_Q$. Vector addition of the Dirac magnetic moments of the quarks then implies:

$$e\mu_p/2M_N = e/2M_Q \tag{7.2a}$$

$$e\mu_n/2M_N = -\frac{2}{3}\, e/2M_Q. \tag{7.2b}$$

Eqs. (7.2a) and (7.2b) yield M_Q = 336 MeV and 328 MeV respectively; the effective binding energy of the nucleon is therefore only 60 MeV. Our use of non-relativistic mechanics is therefore justifiable. The overall picture of the nucleon as a quark-composite is both plausible and consistent.

At a time when the mission of most physicists was to understand why there were 18 low-lying baryons, 8 with spin $\frac{1}{2}$ and 10 with spin $\frac{3}{2}$, and 18 low-lying mesons, 9 with spin 1 and 9 with spin 0, the quark model - with its ability to explain these particles as color-singlet composites of three spin $\frac{1}{2}$ quarks, each of which came in three colors - found ready and universal acceptance, even though many difficulties existed and some remain to this day. Outstanding among these is the problem of setting up a duality principle which will permit one to regard the collective features of hadron dynamics, presumably responsible for the Nambu-Goldstone mechanism, and the "independent-particle" quark model, with its emphasis on free field concepts, as two windows on the same reality. While such problems provide employment for legions of competent physicists attempting to consolidate our gains, let us follow the cutting edge and journey on.

7.3 Composite Leptons and Quarks

We shall restrict ourselves to models suggested by the quark-model paradigm. The constituents shall be taken to be spin - $\frac{1}{2}$ fermions, endowed with an attribute called C" or supercolor, and generically called superquarks; interactions between them are generated by gauging the supercolor degree of freedom and described by a theory labelled QC"D. Leptonic superquarks, the sub-leptons, are to be distinguished from hadronic superquarks, the sub-quarks, until we come to models which seek to provide a unified description of quarks and leptons.

(i) *Leptons*.

Let us focus on the electron and consider a model in which it is deemed to be composed of three spin $\frac{1}{2}$ sub-electrons; the discussion will serve to illuminate a class of problems which arise in composite fermion models.

The first difficulty, which we call the "two length problem", stems from the fact that the electron seems to be a rather point-like object with an enormous Compton wavelength $m_e^{-1} \sim 400$ f. e^+e^- annihilation experiments[5], with PETRA, have shown no evidence of structure down to 2×10^{-3}f. Furthermore, the close agreement between theory and experiment for the parameter $g_e - 2$ may be used to set a sharp upper bound on R_e, the linear extension of the electron[6]. Dimensional analysis tells us that $R_e \sim 1/m^*_{eff.}$, where $m^*_{eff.}$ is the effective sub-electron mass. The Dirac moment of the sub-electron will, in the absence of accidental or other cancellations, contribute terms of order m_eR_e to the gyromagnetic anomaly; hence

$$m_eR_e \sim |a_e^{QED} - a_e^{exp.}| \leqq 5 \times 10^{-10} \qquad (7.3)$$

or

$$R_e \leqq 2 \times 10^{-7} f \qquad (7.4)$$

and

$$m^*_{eff.} \geqq 10^3 \text{TeV} \qquad (7.5)$$

If the effective sub-electron mass is identified with the actual mass, we have a rather absurd situation: 3000 TeV* devoured by binding effects to produce a particle of essentially zero mass. The contrast with the SU(6)-inspired quark-model is striking; compared to this sub-electron model, that model of earlier vintage appears to be the epitome of respectability. Most of us would be tempted to apply Occam's razor and settle for a point electron.

Let us therefore return to the drawing board. A natural way to understand the near masslessness of the electron is to regard it as a consequence of a minutely broken chiral symmetry realized Wigner-Weyl fashion in the electron Hilbert space. [More precisely, in the notation of Lee[4], we consider a type (a) W-W realization which does not accommodate fermion mass; type (b) would allow fermion mass terms at the price of degenerate parity doublets, and would be useless for our purpose. Note that the two types are experimentally distinguishable; if a lepton, ℓ, has an opposite-parity doppelganger: $\sigma(\gamma\gamma\to\ell^+\ell^-) = 2\sigma(\gamma\gamma\to\ell^+\ell^-)_{QED}$. There seems to be no evidence for this factor of 2 for $\ell = e$ or μ.] However, the electron field does not occur in the Lagrangian and the only way to implement chiral sym-

*If terms linear in R_e cancel in the gyromagnetic anomaly, m^*_{eff} can be bounded below by only 23 GeV; the mass that must then be consumed by binding effects need not exceed 70 GeV. For muons this last number is 2 TeV.

metry is to introduce it in the sub-electron Hilbert space. This would mean that $m^* = 0$. Now a light object which is confined to a region of linear extension 10^{-7}f would have kinetic energy equivalent to a mass 2000 TeV. Unlike the situation with the quark model-where we assumed that the flavor chiral group chose the Nambu-Goldstone mode, and noted the possibility of identifying the resultant dynamical mass with the effective confinement mass - the sub-electron mass is exactly zero in the chiral limit. The precise question which urgently calls for an answer may therefore be stated as follows: Can QC"D, a QCD-like theory, confine at distances small enough to keep the electron effectively pointlike and choose the Wigner-Weyl mode for the global chiral group in flavor space?

If the answer to this question is "No", we can close the book on the subject of composite leptons. On the other hand there may exist domains of the coupling constant, or critical values of color and flavor numbers[7], at which the answer is "Yes", in view of this possibility it does not seem inappropriate to take a look at the brighter side of the coin.

The first problem facing any sub-electron model with 3 (or more) spin $\frac{1}{2}$ constituents is the absence of spin $\frac{3}{2}$ states of mass comparable to m_e, indeed, to date, none at any mass-value accessible to present day machines have been seen. (Recall that the first triumph of the quark model and/or SU(6) was an explanation of why there were 8 spin $\frac{1}{2}$ baryons and 10 with spin $\frac{3}{2}$.) The Wigner-Weyl realization of the chiral group, taken in conjunction with a theorem which has long been part of the folklore of physics, provides an explanation. This theorem[8], a consequence of Lorentz invariance which we call the Durand-Weinberg-Witten theorem, states that "massless particles with spin > $\frac{1}{2}$ cannot carry a conserved internal** charge". Hence, barring possible flavor-singlet neutrals, particles with spin > $\frac{1}{2}$ can not exist in the chiral limit; any charge-bearing spin $\frac{3}{2}$ leptons, for example, which may exist would have to be massive enough to have lost all memory of the symmetry limit.

(ii) *Quarks*:

All the salient features of our discussion of sub-electrons are applicable to sub-quarks. Some differences, which are obvious but nonetheless worth noting, are mentioned below. For the moment we restrict ourselves to u and d quarks.

While all existing experiments indicate that quarks appear point-like to leptonic probes - in both the space-like (deep inelastic scattering) and time-like (e^+e^- - annihilation) momentum transfer regimes - there is no independent handle, such as (g - 2), on the quark structure. Indeed since lepton probes are necessary, the upper limits on quark radii are necessarily inferior to the ones for lepton radii. Until the next generation of high energy accelerators goes

**"Conserved internal" means that the charge Q is expressible as $Q = \int d^3x J_o(x,t)$ with $\partial^\mu J_\mu = 0$, where J_μ transforms as a 4-vector under the homogeneous Lorentz group. Massless Yang-Mills fields therefore evade the theorem and may bear conserved charges.

into operation, the best one can say is that

$$R_{quark} < (10^{-3}-10^{-2})f \tag{7.6}$$

or

$$(R_{quark})^{-1} > (10-100)\text{GeV} \tag{7.7}$$

The basic theoretical difference between leptons and quarks lies in the different internal alignments of supercolor and color spins and this has interesting consequences. Leptons are supercolor as well as color singlets; this makes it possible to avoid low-lying spin $\frac{3}{2}$ excitations, in some models, by simply requiring the sub-leptons to obey Fermi statistics. Quarks, on the other hand, are supercolor singlet but color triplets; it is necessary therefore to invoke the theorem of Durand et al to remove potential low-lying spin $> \frac{1}{2}$ excitations, and the caveats associated with this theorem must be borne in mind.

7.4 Hyperfermions as part of a Subquark Multiplet

No other mechanism for generation of current mass (Fig. 6.1), which can be incorporated in a painless way into the hypercolor scenario, has been devised as yet. Recent attempts to generate mass hierarchies in the framework of composite models, notably by Weinberg[9], can not accommodate spontaneous symmetry breaking with an order parameter ≈ 300 GeV; such symmetry breaking is, of course, essential for the hypercolor scenario. (The terminology of Weinberg's paper may confuse the reader; the concept described as "hypercolor" by Weinberg is what we have called "supercolor").

It would be attractive indeed if one could simplify life by identifying hyperquarks with superquarks. However, as noted earlier, such an identification does not seem possible within the framework of current theoretical ideas. Unlike hyperquarks, superquarks do not have a Nambu-Goldstone mass whose presence is heralded by Goldstone bosons. Thus hyperquarks add a new dimension to the problem of quark-proliferation.

To conclude: while the final veredict can not be said to be in, the prospects of composite models of quarks and leptons, as of composite Higgses, do not appear to be terribly bright. Consideration of radically new theoretical ideas is perhaps necessary, if the quest for simplicity, and freedom from the problems of elementary scalars, is to find fulfillment.

8. OUTLOOK

We have discussed modern gauge theories, focusing on the part that appears to be least understood, to wit: the scalar sector. The principal lesson to be learned from our discussion is that things do not, in all likelihood can not, fall neatly into place if one adheres to the logic of Quantum Field Theory in (1+3) - dimensional space. This appraisal of the subject is not pessimistic; it is merely realistic.

Even if the aesthetic unattractiveness of the canonical methodology be overlooked, it is hard to see how the symmetry-breaking process can avoid the trap of triviality; consequently, electroweak theory and grand unification have to be treated as different levels of phenomenology to be used until a better theory -- which embeds the latter and has a wider domain of applicability -- is in hand. We are thus led to progressive unification[1], as the best means available for achieving a deeper understanding of the basic forces; alternatives proposed, in the conventional framework, leave much to be desired.

If there are but four forces, the existence of a satisfactory gravity-theory emerges as the _conditio sine qua non_ for further progress. The superstring concepts, which originated in the work of Green and Schwarz[2], and have been developed by Gross, Witten and others[3], appear to be capable of -- indeed well suited for -- handling gravity. However, the natural mass scale in these formulations is 10^{19} GeV; the problem we face, as indicated in the Introduction, is thus one of descending safely to the electroweak level of 10^{2} GeV rather than of ascent to a unified theory at the Planck mass. When a safe route has been charted, we may have an answer to the following questions: What aspects of the gauge theoretic wisdom are relevant? How much of it will survive the test of time?

9. REFERENCES

Section 1

1. For a review, see: M.A.B. Bég and A. Sirlin, Phys. Reps. 88, 1 (1982). An update of some topics, with recent references, is contained in: M.A.B. Bég, "Scalar Sector of Gauge Theories and the Quest for a Unified Theory", Rockefeller University Report No. RU85/B/128 [to be published in the Proceedings of the International Symposium on Particles and the Universe, Thessaloniki, Greece (1985)].
2. S. Weinberg, Phys. Rev. Lett. 19, 1264 (1967). A Salam, Proc. of Eighth Nobel Symposium (John Wiley, New York, 1968).
3. H.D. Politzer, Phys. Rev. Lett. 30, 1346 (1973); D.J. Gross and F. Wilczek, ibid 30, 1343 (1973).
4. O. Klein, in: New Theories in Physics, Proc. Warsaw Conf. sponsored by the Intern. Institute of Intellectual Cooperation of the League of Nations, Paris (1938) p. 77.
5. C.N. Yang and R.L. Mills, Phys. Rev. 96, 191 (1954).
6. F. Gürsey and L. Radicati, Phys. Rev. Lett. 13, 173 (1964).
7. M.A.B. Bég and V. Singh, Phys. Rev. Lett. 13, 418, 681(E) (1964).
8. M.A.B. Bég, B.W. Lee and A. Pais, Phys. Rev. Lett. 13, 514 (1964).

9. O.W. Greenberg, Phys. Rev. Lett. 13, 598 (1964).
10. For reviews see: H.D. Politzer, Phys. Reps. 14C, 129 (1979); W. Marciano and H. Pagels, ibid 36C, 137 (1978); A. Mueller, ibid 73C, 237 (1981).
11. P.W. Higgs, Phys. Lett. 12, 132 (1964).
12. T.W.B. Kibble, Phys. Rev. 155, 1554 (1967).
13. J.H. Christenson, J.W. Cronin, V.L. Fitch and R. Turlay, Phys. Rev. Lett. 13, 138 (1964).
14. A.D. Sakharov, Pis'ma Zh. Eksp. Teor. Fiz. 5, 32 (1967) [JETP Lett. 5, 24 (1967)] and Zh. Eksp. Teor. Fiz. 76, 1172 (1979) [JETP 49, 594 (1979)].
15. H. Georgi and S.L. Glashow, Phys. Rev. Lett. 32, 438 (1974).
16. H. Georgi, H. Quinn and S. Weinberg, Phys. Rev. Lett. 33, 451 (1974).
17. E.M. Lipmanov, Yad. Fiz. 6, 541 (1967) [Sov. J. Nucl. Phys. 6, 395 (1968)] .
18. M.A.B. Bég, R. Budny, R. Mohapatra and A. Sirlin, Phys. Rev. Lett. 38, 1252 (1977).
19. L.D. Landau, in: Nields Bohr and the Development of Physics (McGraw Hill, New York, 1955).
20. See. for example, M.A.B. Bég, Ref. 1 and articles cited therein.
21. M.A.B. Bég and A. Sirlin, Ann. Rev. of Nuclear Science 24, 379 (1974).
22. For reviews, see: M.A.B. Bég and A. Sirlin, Ref. 1; E. Farhi and L. Susskind, Phys. Reps. 74, 277 (1981).
23. For a "mini review", with references to the recent literature, see: J.H. Schwarz, Comments on Nucl. and Part. Phys. 15, 9 (1985).

SECTION 2.

1. E.S. Abers, and B.W. Lee, Phys. Report 9C, 1 (1973).
2. M.A.B. Bég and A. Sirlin, Phys. Report 88, 1 (1982).
3. S. Weinberg, Phys. Rev. Lett. 19, 1264 (1967).
A. Salam. Proc. Eignth Novel Symp. (John Wiley, New York, 1968).
4. S.L. Glashow, Nucl. Phys. 22, 579 (1961).
A. Salam and J.C. Ward, Phys. Lett. 13, 168 (1964).
5. Reviews relevant to the material of this part include:
E.S. Abers and B.W. Lee, Ref. 1;
M.A.B. Bég and A. Sirlin, Ann. Rev. of Nuclear Science 24, 379 (1974);
S. Weinberg, Rev. Mod. Phys. 46, 255 (1974);
H. Fritzsch and P. Minkowski, Phys. Reports 73, 67 (1981);
J.C. Taylor, Gauge Theories of Weak Interactions (Cambridge Univ. Press, New York, 1976).
T.D. Lee. Particle Physics and Introduction to Field Theory (Harwood Academic Publishers, New York, 1981).
6. For reviews of QCD see, for example:
H.D. Politzer, Phys. Reports. 14C, 129 (1974);
W.J. Marciano and H. Pagels, Phys. Reports 36C, 137 (1978);
A.J. Buras, Rev. Mod. Phys. 52, 199 (1980);

A. Mueller, Phys. Reports 73C, 237 (1981).
7. B.W. Lee, in: Proc. XVI Intern. Conf. on High Energy Physics, eds. J.D. Jackson and A. Roberts (Fermi National Laboratory, Batavia, Il., 1972) Vol. IV, p. 266.
8. M. Kobayashi and T. Maskawa, Prog. Theor. Phys. 49, 652 (1973).
9. L.L. Chau, Phys. Reports 95, 1 (1983).
10. M. Veltman, Nucl. Phys. B123, 89 (1977); Phys. Lett. 91B, 95 (1980); M.S. Chanowitz, M.A. Furman, and I. Hinchliffe, ibid. 78B, 285 (1978); Nucl. Phys. B153, 402 (1979); F. Antonelli, M. Consoli, and G. Corbo, Phys. Lett 91B, 90 (1980).
11. R. Lytel, Phys. Rev. D22, 505 (1980).
12. S.L. Adler, Phys. Rev. 177, 2426 (1969); S. Okubo, Phys. Rev. 179, 1629 (1969).
13. G. 't Hooft, Phys. Rev. Lett. 37, 8 (1976) and Phys. Rev. D14, 3432 (1976).
14. G. Feinberg, Phys. Rev. B140, 1402 (1965).
15. M.A.B. Bég. Phys. Rev. D4, 3810 (1971).
16. N.F. Ramsey, Phys. Reports 43C, 409 (1978)
17. J.S. Bell and R. Jackiw, Nuovo Cimento 60,47 (1969); S.L. Adler, Phys. Rev. 177, 2426 (1969).
18. K. Wilson, Phys. Rev. 179, 1499 (1969).
19. A.A. Belavin, A.M. Polyakov, A.S. Schwartz and Yu. S. Tyupkin, Phys. Lett. 59B, 85 (1975); R. Jackiw and C. Rebbi, Phys. Rev. Lett. 37, 172 (1976); C. Callan, R. Dashen and D. Gross, Phys. Lett. 63B, 334 (1976).

SECTION 3.

1. M.A.B. Bég and A. Sirlin, Section 2, Ref. 2.
2. K.G. Wilson, Phys. Rev. B4, 3184 (1971); K.G. Wilson and J. Kogut, Phys. Reports 12C, 78 (1974) See also: G. Parisi, Nucl. Phys. B100, 368 (1975).
3. That the CP property may provide a means of identifying a true Higgs has been noted by A. Ali and M.A.B. Bég, Phys. Lett. 103B, 376 (1981).
4. M.A.B. Bég, C. Panagiotakopoulos and A. Sirlin, Phys. Rev. Lett. 52, 883 (1984).
5. K. Osterwalder and R. Schrader, Commun. Math. Phys. 42, 281 (1975); See also: V. Glaser, ibid. 37, 257 (1974).
6. J. Glimm and A. Jaffe, Phys. Rev. Lett. 33, 440 (1974).
7. J. Glimm and A. Jaffe, Ann. Inst. Henri Poincaré 22, 97 (1975).
8. M. Aizenman, Phys. Rev. Lett. 47, 1 (1981).
9. J. Frohlich, Nucl. Phys. B200 [FS4], 281 (1982).
10. Wilson and Kogut (Ref. 2); G.A. Baker, Jr. and J. Kincaid, Phys. Rev. Lett. 42, 1431 (1979); J. Stat. Phys. 24, 469 (1981); G.A. Baker, Jr., L.P. Benofy, F. Cooper, and D. Preston, Nucl. Phys. B210, 273 (1982); C.M. Bender, F. Cooper, G.S. Guralnik, R. Roskies, and D.H. Sharp, Phys. Rev. D23, 2976 (1981); 24, 2272 (E) (1982); B. Freedman, P. Smolensky, and D. Weingarten, Phys. Lett. 113B, 481 (1982).
11. W.A. Bardeen and M. Moshe, Phys. Rev. D28, 1372 (1983), and refe-

rences cited therein.
12. M.A.B. Bég and R. C. Furlong, Phys. Rev. D31, 1370 (1985).
13. D.J. Gross and F. Wilczek, Phys. Rev. D8, 3633 (1973); See also: T.P. Cheng et. al., Phys. Rev. D9, 2259 (1974).
14. M.A.B. Bég and A. Sirlin, Ann. Rev. Nucl. Sci. 24, 379 (1974).
15. The relevant β functions may be gleaned from the paper of D.J. Gross and F. Wilczek, Ref. 13 and T.P. Cheng et al., Ref. 13.
16. See, for example, E.L. Ince, Ordinary Differential Equations (Dover, New York, 1956).
17. Almost all bounds, including the ones quoted here, are based on some formulation of the principle that weak interactions lend themselves to a perturbative treatment. See B.W. Lee, C. Quigg, and H.B. Thacker, Phys. Rev. D16, 1519 (1977); N. Cabibbo, L. Maiani, G. Parisi, and R. Petronzio, Nucl. Phys. B158, 295 (1979).
18. K.S. Babu and E. Ma, University of Hawaii Preprint (1985).
19. A. Bovier and D. Wyler, Zurich ETH Preprint (1984).
20. K.S. Babu and E. Ma, Phys. Rev. D31, 2861 (1985) and University of Hawaii Preprint (1984).
21. L.D. Landau, in Niels Bohr and the Development of Physics (McGraw Hill, New York, 1955). See also: N.N. Bogoliubov and D.V. Shirkov, Introduction to the theory of Quantized Fields (Interscience, New York, 1959), Sect. 43.2, and the discussion in Section 1.
22. B.W. Lee, C. Quigg and H.B. Thacker, Ref. 17; N. Cabibbo, L. Maiani, G. Parisi, and R. Petronzio, Ref. 17; R. Dashen and H. Neuberger, Phys. Rev. Lett. 50, 1897 (1893); D.J.E. Callaway, CERN preprint TH. 3360 (1983). [The viewpoint in the last paper seems, at least superficially, similar to ours; the analysis, however, is very different and the results differ.]
23. For a review of the subject, and extensive references to the literature, see: M.A.B. Bég and A. Sirlin, Ref. 1.

SECTION 4.

1. K. Wilson, unpublished remark, cited by L. Susskind, Phys. Rev. D20, 2619 (1979).
2. V.F. Weisskopf, Zeits. fur Phys. 89, 27 (1934); 90, 817 (1934).
3. A. Salam, in: Nonpolynomial Lagrangians, Renormalization and Gravity, Proc. 1971 Coral Gables Conf., eds. M. Dal Cin, G.J. Iverson and A. Perlmutter (Gordon and Breach, New York, 1971) p. 3 and references cited therein.
4. H. Georgi and C. Jarlskog, Phys. Lett. 86B, 297 (1979).
5. M. Wise, H. Georgi and S.L. Glashow, Phys. Rev. Lett. 47, 402 (1981).
6. F. Gürsey (private communication).
7. S.L. Glashow (public communication).

SECTION 5.

1. See, for example, E. Farhi and R. Jackiw, "Dynamical Gauge Symmetry Breaking" (World Scientific, Singapore, 1982).
2. M.A.B. Bég and A. Sirlin, Ann. Rev. of Nuclear Science 24, 379 (1974).[See, especially, the last paragraph of Section 7.2].
3. M.A.B. Bég, in New Frontiers in High Energy Physics, Proceedings of Orbis Scientiae 1978, edited by A. Perlmutter and L. Scott (Plenum, New York, 1978).
4. S. Weinberg, Phys. Rev. D13, 974 (1976).
5. L. Susskind, SLAC Report No. 2142, 1978 (Phys. Rev. D, to be published).
6. P.W. Anderson, Phys. Rev. 112, 1900 (1958).
7. J. Schwinger, Phys. Rev. 128, 2425 (1962).
8. J.M. Cornwall and R.E. Norton, Phys. Rev. D8, 3338 (1973).
9. R. Jackiw and K. Johnson, Phys. Rev. D8, 2386 (1973).
10. S. Coleman, Comm. in Math. Phys. 31, 259 (1973).
11. Cf. L. Dolan and R. Jackiw, Phys. Rev. D9, 3320 (1974).
12. M.A.B. Bég and S.S. Shei, Phys. Rev. D12, 3092 (1975).
13. See, for example, D.J. Gross and A. Neveu, Phys. Rev. D10, 3235 (1974).
14. C.G. Callan Jr., R. Dashen and D.J. Gross, Phys. Rev. D17, 2717 (1978); D. Caldi, Phys. Rev. Lett. 39, 121 (1977); R.D. Carlitz, Phys. Rev. D17, 3225 (1978).
15. G. 't Hooft, Phys. Rev. Lett. 37, 8 (1976).
16. The apt terminology is due to M. Gell-Mann, (Private communication).
17. Constituent u and d masses ~300 MeV may be inferred from quark-model calculations of the absolute values of proton and neutron magnetic moments [M.A.B. Bég, B.W. Lee and A. Pais, Phys. Rev. Lett. 13, 514 (1964)] if the quarks are presumed to have no anomalous or Pauli magnetic moment. For two very different procedures for assessing the mass scale in QCD see: M.A.B. Bég, Phys. Rev. D11, 1165 (1975) and H. Georgi and H.D. Politzer, ibid. D14, 1829 (1976).

SECTION 6.

1. The possibility of getting by without new colors, by introducing exotic quarks belonging to higher (6, 8, 10, ... etc.) representations of $SU(3)_C$ has been discussed by W. Marciano, Rockefeller University Report No. COO-2232 B-193 (1980).
2. S. Weinberg, section 5, Ref. 4.
3. M.A.B. Bég, H.D., Politzer and P. Ramond, Phys. Rev. Lett. 43, 1701 (1979).
4. M.A.B. Bég and A. Sirlin, Section 5, Ref. 2.
5. This horizontal group was labelled $SU(2)_k$ in: M.A.B. Bég, Phys. Rev. D8, 664 (1973). This paper appears to have been the first to discuss the possibility that interactions generated by gaug-

ing this group, the so-called "horizontal interactions", may be of physical relevance. For other discussions of such interactions, see: M.A.B. Bég and A. Sirlin, Phys. Rev. Lett. 38, 1113 (1977); M.A.B. Bég and H.-S. Tsao, ibid. 41, 279 (1978); S. Barr and A. Zee, Phys. Rev. D17, 1854 (1978); F. Wilczek and A. Zee, Phys. Rev. Lett. 42, 421 (1979).
6. S. Dimopoulos and L. Susskind, Nucl. Phys. B155, 237 (1979).
7. One of the problems with EHC is: How does the symmetry breakdown of the EHC group take place at about 30 TeV (or whatever may be the appropriate scale)?. S. Raby, S. Dimopoulos and L. Susskind Nucl. Phys. B169, 373 (1980) have suggested that a large EHC - group may naturally break down in a hierarchy of steps (Self-break or "Tumble"). See also: R.K. Kaul, Rev. Mod. Phys. 55, 449 (1983).
8. S. Dimopoulos and J. Ellis, CERN Preprint TH. 2949 (1980).
9. S. Dimopoulos and L. Susskind, Ref. 6. See also: E. Eichten and K.D. Lane, Phys. Lett. B90, 125 (1980).
10. S. Dimopoulos, H. Georgi and S. Raby, Phys. Lett. 127B, 101 (1983).
11. M.A.B. Bég, Phys. Lett. 124B, 403 (1983).
12. S. Dimopoulos and J. Ellis, Nucl. Phys. B182, 505 (1981).
13. A. Salam, Proc. XXIth Intern. Conf. on High Energy Physics (Paris, 1982).
14. P. Fayet, Proc. XXIth Intern. Conf. on High Energy Physics (Paris, 1982).
15. E. Farhi and L. Susskind, Phys. Rep. 74, 277 (1981)
16. M.A.B. Bég and A. Sirlin, Section 2, Ref. 2.
17. Cf. L. -F. Li, Phys. Rev. D9, 1723 (1974)
18. M.A.B. Bég. Phys. Lett. 129B, 113 (1983).
19. M.A.B. Bég and S.S. Shei, Phys. Rev. D12, 3092 (1975).
20. H.D. Politzer, Nucl. Phys. B117, 397 (1976).
21. M.K. Gaillard and B.W. Lee, Phys. Rev. D10, 897 (1974).
22. Cf. E. Fermi, Ric. Sci. VII-II, 13 (1936).
23. T. Appelquist, M.J. Bowick, E. Cohler, and A.I. Hauser, Phys. Rev. D31, 1676 (1984).
24. A. Zepeda, Phys. Lett. 132B, 407 (1983).
25. M.A.B. Bég in "Recent Developments in High Energy Physics", Edited by B. Kursunoglu, A. Perlmutter and L.F. Scott (Plenum Publishing Corporation, New York, 1980) p. 23.
26. M.A.B. Bég, Proc. XXth Intern. Conf. on High Energy Physics, Madison, eds. L. Durand and L. Pondrom (American Institute of Physics, New York, 1980) p. 489.
27. S. Weinberg, Phys. Rev. Lett. 29, 1698 (1972).
28. S. Dimopoulos and L. Susskind, Ref. 6.
M.A.B. Bég, Ref. 26
E. Eichten and K. Lane, Ref. 9
29. M.A.B. Bég, Refs. 25, 26; Proc. VPI Workshop on Weak Int. as Probes of Unif., eds. G.B. Collins, L.N. Chang and J.R. Ficenec (Am. Inst. of Phys., New York, 1981) p. 505; M.A.B. Bég, in: Proc. of EPS International Conf. on High Energy Physics, Lisbon, Portugal (1981). This report contains an extensive list of references.

30. S. Dimopoulos, S. Raby and G.L. Kane, Nucl. Phys. B182, 77 (1981); See also: F. Hayot and O. Napoly, C.E.N. Saclay preprint Dph-T-86- (1980).
31. A. Ali, H.B. Newman and R.Y. Zhu, Nucl. Phys. B191, 93 (1981).
32. A. Ali and M.A.B. Bég, Phys. Lett. 103B, 376 (1981).
33. J. Ellis, M. Gaillard, D. Nanopoulos and P. Sikivie, Nucl. Phys. B182, 529 (1981).
34. L.L. Chau Wang, BNL Report No. 28781-R (1980).

SECTION 7.

1. O.W. Greenberg and C.A. Nelson, University of Maryland Technical Report No. 74-006 (1973) (unpublished) and Phys. Rev. D10, 2567 (1974);
O.W. Greenberg and G.B. Yodh, Phys. Rev. Lett. 32, 1473 (1974);
O.W. Greenberg, Phys. Rev. Lett. 35, 1120 (1975);
O.W. Greenberg and J. Sucher, Phys. Lett. B99, 339 (1981);
J.C. Pati and A. Salam, Phys. Rev. D10 275 (1974);
J.C. Pati, A. Salam and J. Strathdee, Phys. Lett. 58B, 265 (1975);
H. Terazawa, Y. Chikashige and K. Akama, Phys. Rev. D15, 480 (1977);
H. Terazawa, Prog. Theoret. Phys. 64, 1763 (1980);
H. Harari, Phys. Lett. 86B, 83 (1979).
H. Harari and N. Seiberg, Phys. Lett. 98B, 269 (1981);
M.A. Shupe, Phys. Lett. 86B, 87 (1979);
E.J. Squires, Phys. Lett. 94B, 54 (1980);
C.A. Nelson, Phys. Lett. 93B, 143 (1980);
Y. Ne'eman, Phys. Lett. 82B, 69 (1979);
R. Casalbuoni and R. Gatto, Phys. Lett. 90B, 81 (1980);
M. Yasue, Phys. Lett. 91B, 85 (1980);
J. Ellis, M.K. Gaillard and B. Zumino, Phys. Lett. 94B, 343 (1980);
V. Visnjic-Triantafillou, Phys. Lett. 95B, 47 (1980);
E. Derman, Phys. Lett. 95B, 369 (1980);
I. Bars and S. Yankielowicz, Phys. Lett. 101B, 159 (1981);
2. M.A.B. Bég, B.W. Lee and A. Pais, Phys. Rev. Lett. 13, 514 (1964).
3. F. Gürsey and L.A. Radicati, Phys. Rev. Lett. 13, 173 (1964).
4. B.W. Lee, Chiral Dynamics (Gordon and Breach Science Publishers, New York, 1972);
M.A.B. Bég, Lectures delivered at Centro de Investigación del I.P.N. (México, 1971);
H.R. Pagels, Phys. Reports 16C, 220 (1975).
5. G. Wolf, DESY Report No. 80/13 (1980).
6. S.S. Brodsky and S.D. Drell, Phys. Rev. D22, 2236 (1980);
G.L. Shaw, D. Silverman and R. Slansky, Phys. Lett. 94B, 57 (1980);
Cf. H.J. Lipkin, Phys. Lett. 89B, 358 (1980).
7. J. Preskill and S. Weinberg, Phys. Rev. D24, 1059 (1981).
8. L. Durand III, Phys. Rev. 128, 434 (1962);
S. Weinberg and E. Witten, Phys. Lett. 96B, 59 (1980).

9. S. Weinberg, University of Texas preprint (1981).

SECTION 8.

1. M.A.B. Bég, Rockefeller University Report No. RU 85/B/128 (to be published in the Proceedings of the International Symposium on Particles and the Universe, Thessaloniki, Greece, 1985).
2. See, for example, J. Schwarz, Comments on Nucl. and Part. Phys. 15, 9 (1985) and references cited therein.
3. D. Gross, J. Harvey, E. Martinec and R. Rohm, Phys. Rev. Lett. 54, 502 (1985); P. Candelas, G. Horwitz, A. Strominger and E. Witten, Nucl. Phys. B258, 46 (1985).

HADRON COLLIDERS, INTERMEDIATE VECTOR BOSONS, NEW QUARKS AND LEPTONS*

DAVID B. CLINE
University of Wisconsin, Madison, WI 53706

ABSTRACT

In these lectures we briefly review the current generation of hadron colliders ($\bar{p}p$ at CERN and FNAL) and the experiments at these machines. We also describe the observation and properties of the Intermediate Vector Bosons. Finally we discuss the production of b quarks, the probable observation of the t quark and the search for the next generation of leptons and the observation of dimuon events at the $S\bar{p}pS$ collider.

CONTENTS

1. Antiproton-Proton Colliders
2. Detectors UA1 and CDF
3. Intermediate Vector Bosons and High Energy Electro Weak Parameters
4. The Search for New Quarks-Evidence for the t Quark from $W \to t\bar{b}$
5. The Search for New Leptons: $W \to L + \nu_L$
6. Multimuon Final States and $B\bar{B}$ Production
7. Acknowledgements

1. Antiproton-Proton Colliders

The initial proposal to convert the CERN and FNAL proton synchtrotrons to $p\bar{p}$ colliders was made in 1976 by Rubbia et al.[1] The basic idea was to add an antiproton source to the machine and to store and collide p and $\bar{p}$ at the highest possible energy. The CERN machine started operation in 1981 and the FNAL TeVI $\bar{p}p$ collider is to start operation in 1985. A brief summary of the parameters of these machines is given in Tables 1 and 2.

Because the low-beta value of 1 m is larger than the rms bunch length (0.4 m), a formula for luminosity valid for constant cross section is a good approximation. Thus

$$L = \frac{N_p \, N_{\bar{p}} \, B \, f_o}{4\pi\sigma^2},$$

where $N_p = 10^{11}$ is the number of protons per bunch, $N_{\bar{p}} = 8 \times 10^{10}$, the number of antiprotons per bunch, $B = 3$, the number of bunches per beam, $f_o = 4.77 \times 10^4$ Hz, the revolution frequency. This gives

---

Invited lectures at EMPC, Mexico, December, 1984.

$$L = 2.2 \times 10^{30}\ cm^{-2}\ sec^{-1}.$$

for the FNAL (TeVI) $\bar{p}p$ collider.

Higher luminosity figures can also be expected, as high as $6 \times 10^{30}\ cm^{-2}\ sec^{-1}$, by either improving the stochastic cooling rate (a factor of two), or the longitudinal beam density in the Main Ring and Tevatron (also possibly by a factor of two) or with 6 bunches per beam instead of 3 as assumed here.

References (4)

1. C. Rubbia, P. McIntyre, D. Cline, Proceedings of the Aachen Neutrino Conference, 1976.
2. FNAL Antiproton Source Design Report, Unpublished (1982).

2. Detector (UA1 and CDF)

We will review two specific detectors that illustrate different approaches to the detection of particles in $\bar{p}p$ collisions. (1) The UA1 experiment at CERN (Sp$\bar{p}$S collider and (2) the CDF detector at FNAL (TeVI $\bar{p}p$ collider). The full compliment of detectors at CERN is given in Table 3

The basic particle identification techniques are given in Table 4.

The UA1 detector employs a large conventional dipole magnetic field. A drawing of the detector is shown in Fig. 1. The various components of the calorimeter detection are given in Table 5.

The CDF detector employs a very large superconducting solenoid magnetic field. The detector is divided into components for the central region and for the forward region. This is necessary due to the larger rapidity range for the TeVI $\bar{p}p$ collider. A schematic drawing of the CDF detector is given in Fig. 2.

Table 1

AA Ring at CERN

$\bar{p}$ momentum	3.5 GeV/c	
Mean Radius	25 m	
Regular Lattice with 2 high, 2 zero dispersion regions		
Betatron Tunes	~2.4	
Cooling Systems		Band Width
System		
Momentum precooling (before stock)		400 MHz
Stack Tail (Accumulation) System		400 MHz
Stack Betatron systems		400 MH
Core Momentum		1-2 GHz
Core Betatron		1-2 GHz
$\bar{p}$ parameters, acceptance from Target		
Emittance		$75\pi\mu m$
Momentum Spread		1.5%
Number ~		$5x10^6$
Horn Collection system		

$\bar{p}$'s are precooled in momentum, then RF stacked in tail

$L_{pp} = 5x10^{29}$ cm^{-2} sm^{-1} for $N\bar{p} = 2 x10^{11}$

Table 2

Antiproton Stack Parameters for the Fermilab Antiproton Source

Injected Pulse	
Number of $\bar{p}$'s	7×10^7
$\Delta p/p$	0.2%
Horizontal and vertical emittance	10π mm-mrad
Time between injections	2 sec
Fraction of beam accepted	>85% of injected pulse
Flux	3×10^7 $\bar{p}$/sec
Final Stack	
Number of $\bar{p}$'s	4.3×10^{11}
$\Delta p/p$	0.1%
Horizontal and vertical emittance	2π mm-mrad
Peak density	1×10^5 eV^{-1}
Core width (Gaussian part)	1.7 MeV (rms)
Total stacking time	

UA1

LOCATION	CERN SPS $\bar{p}p$ Collider CERN, Geneva, Switzerland
MAGNET	Dipole field up to 0.7 Tesla, Al coils Magnetic volume = $7.0 \times 3.5 \times 3.5$ m^3
CENTRAL DETECTOR	Cylinder (6 m long, 2.2 m diameter) made of 6 independent modules containing drift chambers with 18 cm drift space Covers $5° < \theta < 175°$ 60% ethane - 40% argon 3-dimensional readout by continuous digitization in drift direction and charge division along wires Average of 110 space points per track $\sigma \simeq 250$ μm in drift plane $\sigma \simeq 2$ % of wire length along wire $\sigma \simeq 6\%$ for dE/dx
ELECTROMAGNETIC CALORIMETERS	Pb-scintillator sandwich ($26X_0$) with BBQ readout Gondolas ($25° < \theta < 155°$) and Bouchons ($5° < \theta < 25°$, $155°$ θ $<175°$) $(\sigma_E/E)^2 = (0.15/\sqrt{E})^2 + (0.016)^2$ $\sigma_x = 4$ cm$/\sqrt{E}$ (θ direction) $\sigma_y = 16$ cm$/\sqrt{E}$ (ϕ direction) $\sigma_{E_t}/E_t = 0.12/\sqrt{E_t}$
HADRON CALORIMETERS	Fe-scintillator sandwich with BBQ readout based on the laminated return yoke of the magnet 16 samplings (5 cm Fe, 1 cm scint. each) in barrel, 23 samplings in endcaps $\Delta E/E \simeq 0.8/\sqrt{E}$
MUON DETECTION	Large-area drift tube chambers (8 layers) Angular resolution $\sigma = 1$ mrad
FORWARD DETECTORS	$0.2° < \theta < 5°$ and $0 < \phi < 2\pi$ Rapidity acceptance $3.4 < \|y\| < 7.4$ Endcap chambers Trigger counters (4 cm thick scint.) Electromagnetic calorimeter (4 modules Pb-scint., $7.2X_0$ each) EM shower chambers between first and second EM modules (proportional chambers delay line readout) Hadron calorimeter (6 modules, 1.7 λ_{abs} each), based on compensating magnet steel Hadron shower chamber between first and second hadron calorimeter modules
ROMAN POTS	8 small drift chambers, 4 on each arm at ± 22 m from collision point, which enter the SPS vacuum pipe vertically

REFERENCES

1. G. Arnison et al., Phys. Lett. **107B** (1981) 320.
2. M. Calvetti, The UA1 Central Detector, Proceedings of Int. Conf. on Instrumentation for Colliding Beam Physics, SLAC-250 (Feb. 17-23, 1982).
3. M. Barranco-Luque et al., Nucl. Instr. & Meth. **176** (1980) 175.
4. M. Calvetti et al., Nucl. Instr. & Meth. **176** (1980) 217.

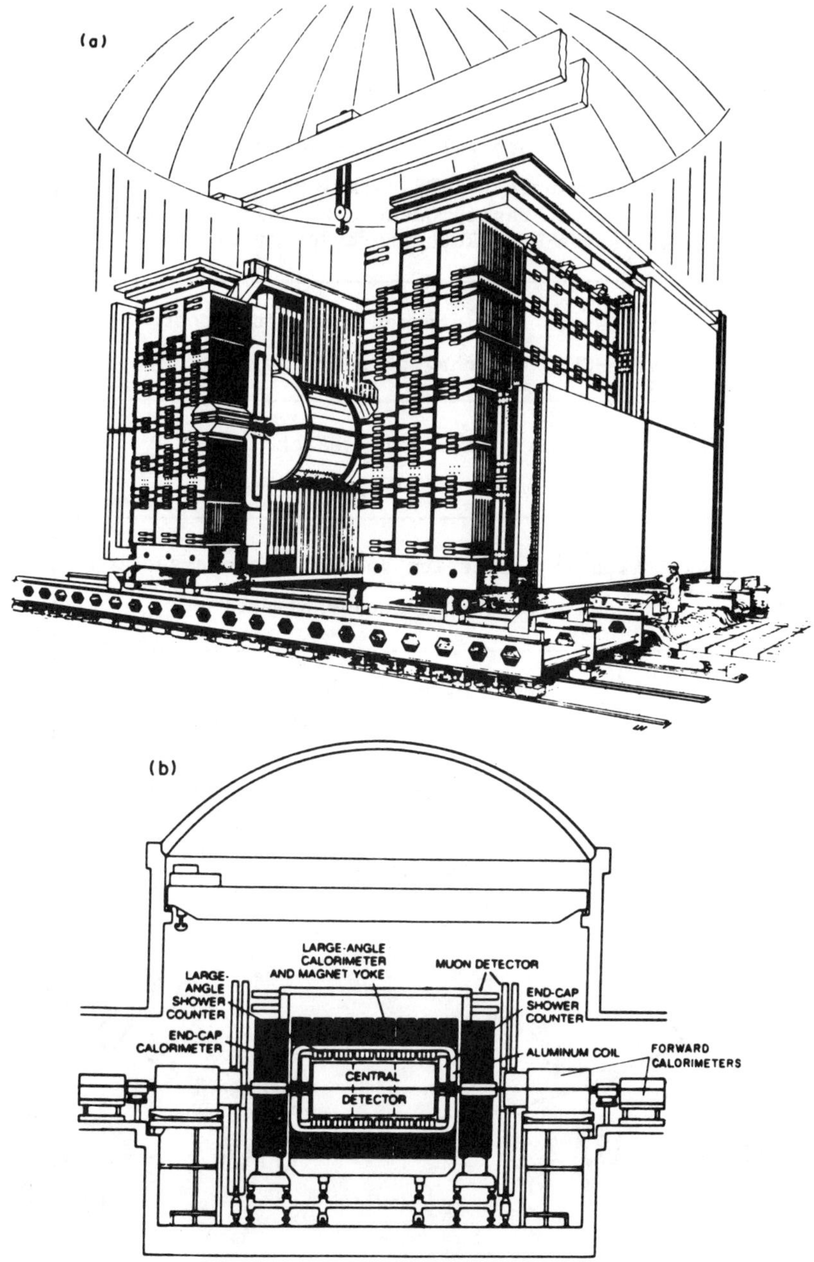
(a)
(b)
LARGE-ANGLE CALORIMETER AND MAGNET YOKE
MUON DETECTOR
LARGE-ANGLE SHOWER COUNTER
END-CAP SHOWER COUNTER
END-CAP CALORIMETER
ALUMINUM COIL
FORWARD CALORIMETERS
CENTRAL DETECTOR

Table 3

$p\bar{p}$ Experiments at CERN

UA1: General-purpose detector Aachen-Annecy (LAPP)-Birmingham-CERN-Paris (Coll. de France)-Queen Mary College, London-Riverside-Rome-Rutherford-Saclay-Vienna
UA2: Search for intermediate vector bosons Berne-CERN-Copenhagen-Orsay-Pavia-Saclay
UA3: Monopole search Annecy (LAPP)-CERN
UA4: Elastic scattering and total cross-section Amsterdam-CERN-Genoa-Naples-Pisa
UA5: Streamer chamber Bonn-Brussels-Cambridge-CERN-Stockholm

LOCATION

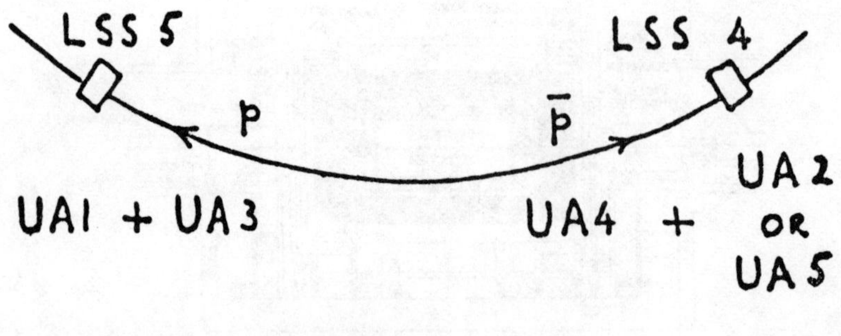

Table 4

Particle Identification at $\bar{p}$ Colliders

	Method
1. Hadron Jet	Energy Deposition in EM and Hadronic Calorimeter
2. Electron	Energy Deposition in EM Calorimeter Only $P_{trad} \approx E_{calor}$ $P_{\perp} > 10$ GeV/c to reduce τ background
3. Neutrino or Photiono...	Missing Transverse Energy > 15 GeV (UA1 only)
4. Muon	Penetrating Particle into Muon Chambers $P_{track_{\perp}} > 5$ GeV/$_c$ - No $K \rightarrow \mu + \nu$ decay
4. K^o/Λ	V_{ee} fit in tracking system with current mass
5. Photon	EM only - good localization (next generation of UA1...)

Table 5

The UA1 Calorimetery

Calorimeter		Angular coverage θ (°)	Thickness: No. rad. lengths	Thickness: No. abs. lengths	Cell size: Δθ (°)	Cell size: Δφ (°)	Sampling step	Segmentation in depth	Resolution
Barrel	e.m.: gondolas	25-155	26/sin θ	1.1/sin θ	5	180	1.2 mm Pb 1.5 mm scint.	3.3/6.6/9.9/6.6 X_0	$0.15/\sqrt{E}$
	hadr.: c's		-	5.0/sin θ	15	18	50 mm Fe 10 mm scint.	2.5/2.5 λ	$0.8/\sqrt{E}$
End-caps	e.m.: bouchons	5-25	27/cos θ	1.1/cos θ	20	11	4 mm Pb 6 mm scint.	4/7/9/7 X_0	$0.12/\sqrt{E_T}$
	hadr.: I's	155-175	-	7.1/cos θ	5	10	50 mm Fe 10 mm scint.	3.5/3.5 λ	$0.8/\sqrt{E}$
Calcom	e.m.	0.7-5	30	1.2	4	45	3 mm Pb 3 mm scint.	4 × 7.5 X_0	$0.15/\sqrt{E}$
	hadr.	175-179.3	-	10.2	-	-	40 mm Fe 8 mm scint.	6 × 1.7 λ	$0.8/\sqrt{E}$
Very forward	e.m.	0.2-0.7	24.5	1.0	0.5	90	3 mm Pb 6 mm scint.	5.7/5.3/5.8/7.7 X_0	$0.15/\sqrt{E}$
	hadr.	179.3-179.8	-	5.7	0.5	90	40 mm Fe 10 mm scint.	5 × 1.25 λ	$0.8/\sqrt{E}$

3. Intermediate Vector Bosons and High Energy Electro Weak Parameters

The discovery of the W and Z particles will make it possible to measure the high energy electroweak parameters ρ and $\sin^2\theta_W$ with great accuracy.[1,2] This will also provide a test of the radiative corrections. In addition the decays of W and Z provide a laboratory for the observation of new particles. The first example is the observation of the t quark through[3]

$$W \to t + \bar{b}$$

In addition the decay

$$Z \to \nu\bar{\nu}$$

provides a sensitive test for the number of neutrinos in nature.[4] Also the decay

$$W \to L + \nu_L$$

into a new heavy lepton could "discover" the next generation.[5] In addition the decay of W into old leptons (e,μ,τ) can be used to test for universality of leptons and might some day lead to a better understanding of the universality of leptons.

Finally, $\bar{p}p$ colliders produce copious numbers of b quarks and we will show that the present UA1 data as same sign dimuons can limit B^o–$\bar{B}^o$ mixing parameters.

In Table 6 we give the definition of the high energy parameters that determine the success of the SU(2)xU(1) theory.

Table 6

High Energy Electroweak Parameters

1. Spin W,Z; (V-A)/(V+A) Components, $\sin^2\theta_W, \rho$
2. $M_W = \left[\dfrac{\pi\alpha}{\sqrt{2}\, G_\mu \sin^2\theta_W (1-\Delta r)} \right]^{1/2}$
3. $M_Z = M_W/\cos\theta_W = \dfrac{77.3 \pm 0.08 \text{ GeV}}{\sin^2\theta_W}$
4. Γ_W, Γ_Z
5. N_ν - number of neutrino families
6. Mass Heavy Sequential Lepton, M_t quark

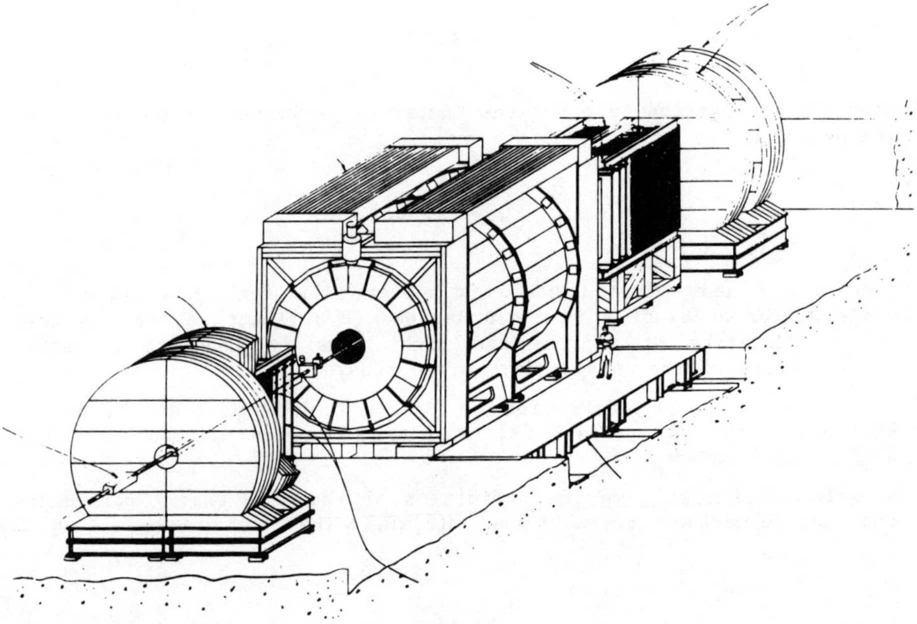

Figure 2

One goal of the extensive improvement program for the UA1 detector to measure these parameters using W and Z particles from the $\bar{p}p$ collider.[6]

The 1983 $\bar{p}p$ collider run at CERN yielded a significant number of W and Z events analyzed by both the UA1 and UA2 groups. During the 1984 run the number of W and Z events has increased by about a factor of 2.5 bringing the total of W and Z events to 0(300)(W) and 0(30)(Z).[7]

The first goal of the study of the W and Z is to identify the largest possible number of expected decay modes. In Tables 7 and 8 we give a summary of the decay channels that have been identified so far.

Table 7

W Decays

Decay Mode	Identification Status
$W \rightarrow e + \nu_e$	Yes
$W \rightarrow \mu + \nu_\mu$	Yes
$W \rightarrow \tau + \nu_\tau$	Yes
$W \rightarrow L + \nu_L$	Limit only
$W \rightarrow u + \bar{d}$	No (indirectly by $u + \bar{d}$, W production)
$W \rightarrow c + \bar{s}$	Most likely
$W \rightarrow t + \bar{b}$	Most likely
$W \rightarrow H^o + e + \nu$ $H^o + \mu + \nu$ $H^o + q + \bar{q}$	No

Table 8

Decays of Z

Decay Mode	Identification Status
$Z \rightarrow e^+e^-$	Yes
$Z \rightarrow \mu^+\mu^-$	Yes
$Z \rightarrow \nu_i\bar{\nu}_i$	Possibly (High $P_\perp$ gluon recoil)
$Z \rightarrow \tau\bar{\tau}$	Yes ($\tau \rightarrow$ hadrons $\bar{\tau} \rightarrow \bar{\ell} + \nu + \nu$)
$Z \rightarrow u + \bar{u}$	No ($u\bar{u} \rightarrow Z^o$ indirectly)
$Z \rightarrow d + \bar{d}$	No ($d\bar{d} \rightarrow Z^o$ indirectly)
$Z \rightarrow s + \bar{s}$	No
$Z \rightarrow c + \bar{c}$	Yes ($c \rightarrow \mu$)
$Z \rightarrow b + \bar{b}$	Yes ($b \rightarrow \mu$)
$Z \rightarrow t + \bar{t}$	No (M_t, Phase space)
Exotic Decays $Z \rightarrow \mu\mu H^o$...	No

The 1983 run parameters of the W and Z particles from the UA1 experiment are given in Table 9. In Table 10 are given preliminary results from the UA2 experiment (1984) run. Note the excellent agreement with the low energy SU(2) x U(1) parameters. The next step will be to make precise measurement of these parameters that test the theory at the loop calculation level and hence the radiative corrections.

Table 9

Measured Properties of IVB's

(UA1)

	$W^{\pm}$	Z^{o}	
Mass	80.9±2.8 GeV/c^2	93.9±2.9 GeV/c^2	
Full Width (90% CL)	< 7 GeV	< 8.5 GeV	
Observed Decays	$e^{\pm}\nu$ (68) $\mu^{\pm}\nu$ (14)	$e^{+}e^{-}$ (4) $\mu^{+}\mu^{-}$ (5)	e: $\sin^2\theta_W = 0.226 \pm 0.008 \pm 0.014$ $\rho = 0.96 \pm 0.05$
$\sigma \bullet B$ $\sqrt{s}$=540GeV $p\bar{p}$	0.53 ± 0.08 nb (± 0.09)	58 ± 21 pb (± 9)	μ: $\sin^2\theta_W = 0.228 {}^{\pm 0.04}_{\ \ 0.03}$ $\rho = 1.15 {}^{+0.22}_{-0.28}$
Spin	1	---	

Table 10

Data from UA2 - 1984 Run (Ref. 8)

$\sigma_Z^e = 0.11 \pm .04 \pm .02$ nb

$\sigma_W{}^e = 0.53 \pm 0.1 \pm .1$ nb

QCD $\Gamma w/\Gamma_Z = (9.3\pm 9)\ \sigma_e/\sigma_W{}^e$

$\Gamma_W = 2.77\ GeV/c^2$

$\Gamma_Z < 2.6 \pm GeV/_c2$ (90%)

$\Gamma_Z < 3.1 \pm 0.3\ GeV/_c2$ (95%)

$N_{Z\to e^+e^-} = 16$ (83 + 84 data)

$N_{W\to e\nu} \approx$ (83 + 84 data)

$M_W = 83.1 \pm 1.9 \pm 1.3\ GeV/_c2$

$M_Z = 92.7 \pm 1.7 \pm 1.4\ GeV/_c2$

A comparison of the parameters measured by the UA1 and UA2 group is given in Table 11 and compared in Figure 1.

Table 11

Comparison (UA1/UA2)

	UA1	UA2	Combined
M_W	80.9±2.8	83±1.9±1.3	81.8±2.6
M_Z	93.9±2.9	92.7±1.7±1.4	93.3±2.7
$\sin^2\theta_W$	0.226±0.016	0.216±0.01±0.007	0.221±0.013
ρ	0.96±0.05	1.02±0.06	0.99±0.05

$$\sin^2\theta_W = 0.221 \pm 0.01$$

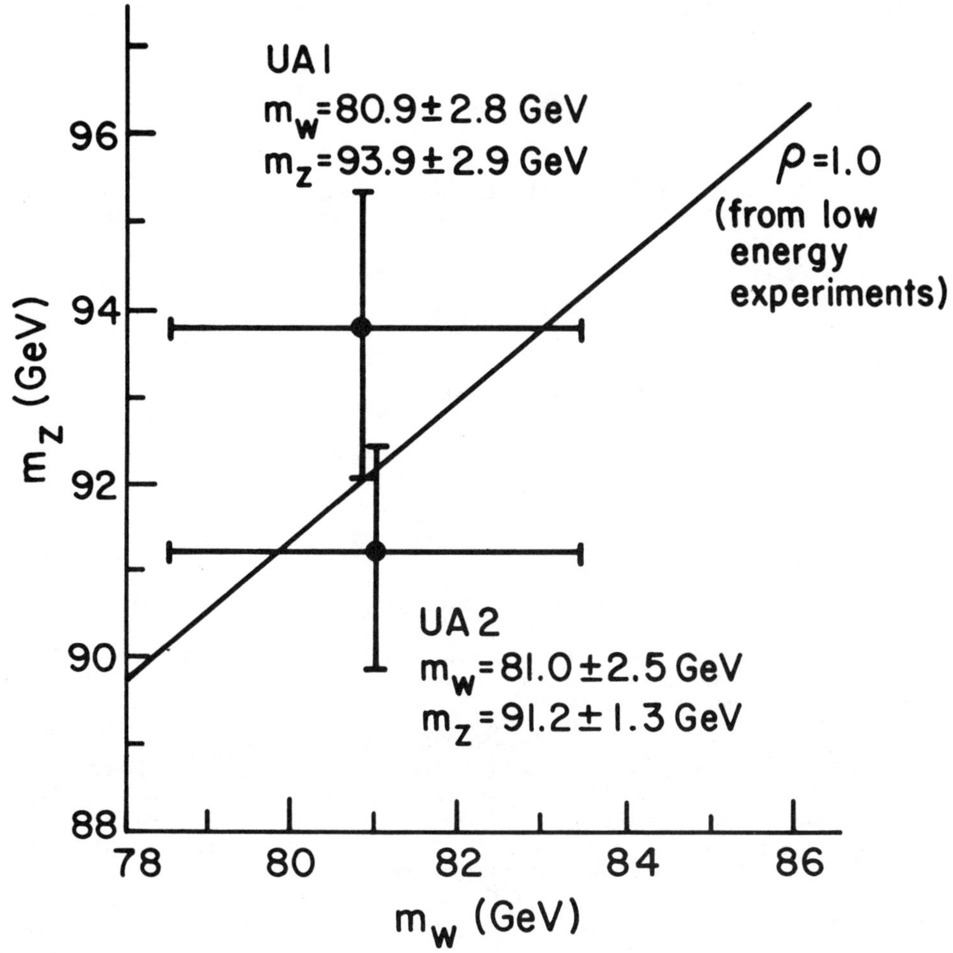
UA1
m_W=80.9±2.8 GeV
m_Z=93.9±2.9 GeV
ρ=1.0
(from low energy experiments)
UA2
m_W=81.0±2.5 GeV
m_Z=91.2±1.3 GeV
96
94
92
90
88
78
80
82
84
86
m_Z (GeV)
m_W (GeV)

Fig. 3

The radiative corrections are expressed as Δr where the value of the G_μ weak coupling constant can be expressed as

$$\frac{G_\mu}{\sqrt{2}} = \frac{g^2}{8M_W^2[1-\Delta r]}$$

where g is the SU(2)xU(1) coupling constant. The radiative correction Δr is given by a sum of terms

$$\Delta r = \Delta r^{(1)} + \Delta r^{(2)} + \Delta r^{(3)}$$

where $\Delta r^{(1)}$ comes from corrections due to gauge bosons, $\Delta r^{(2)}$ due to leptonic effects and $\Delta r^{(3)}$ due to hadronic effects. The best estimate for Δr is

$$\Delta r = 0.0696 \pm 0.002$$

Δr is related to the M_W and M_Z parameters as

$$\Delta r = 1 - \frac{[37.28 \text{ GeV}]^2}{(\frac{M_W2}{M_Z2})(M_Z2-M_W2)}$$

Thus a measurement of M_Z-M_W and $[M_W/M_Z]$ will allow determination of Δr. The theoretical estimate for Δr (or M_Z-M_W) depends on one other parameters (i.e $\sin^2\theta_W$ for example). However the precise absolute measurement of M_Z and a precise measurement of M_Z-M_W are sufficient to tie down the theory completely.

As an example if M_Z = 93.8±0.3 GeV the theoretical values for M_Z-M_W has an uncertainty of ~70 MeV (i.e. once M_Z has been determined to 300 MeV)* Many new effects can cause a larger deviation of M_Z-M_W (Δr). For example in Table 12 (Marciano)[9] is shown the dependence of M_Z-M_W on the t-quark mass. Note M_t from 36-60 GeV causes a shift of 30 MeV. A 1 TeV Higgs boson gives comparable changes.

Table 12

m_t(GeV)	$\Delta\Gamma$	A(GeV)	m_Z-m_W
20	0.0699	38.66	10.79
36	0.0696	38.65	10.79
60	0.0714	38.69	10.82
83	0.0623	38.50	10.86
100	0.0558	38.37	10.56
150	0.0379	38.01	10.27
200	0.0167	37.60	9.96
240	-0.0036	37.21	9.68
Tree level	0	37.28	9.73

Above evaluated for $\sin^2\theta_W = 0.217$

There are theoretical uncertainties in the predicted properties due to the unknown number of lepton and neutrino families; indeed the Z^o decays provides an important source of decay into all neutrino families

$$Z^o \rightarrow {}_i\Sigma(\nu_i + \bar{\nu}_i)$$

and these decays in turn increase the Z^o width by ~180 MeV for each new family. At present three families are known (ν_e, ν_μ, ν_τ). The decays of the OW is less sensitive to the number of neutrinos but will depend on the existence of heavy leptons and heavy quarks (see Figure 4). The 1983 results from the UA1 experiment are shown in Fig. 5. In time we expect to understand the bulk of the W decays and therefore the W width "exactly" calculable. In contrast, the Z^o width is not calculable unless the neutrino decay can be directly observed, a formidable if not impossible task. A simple technique has been proposed to estimate the Z^o width from the production rates of W and Z events and explore the limits of this technique.[10]

We assume that the absolute rates of $W \rightarrow \ell + \nu_\ell$ and $Z \rightarrow \ell + \bar{\ell}$ and the ratio of production cross sections σ_Z/σ_W are described within the

Decay of the Z^0 into Neutrino Families

	Factor in Branching Ratio
$Z^0 \to \nu_e \bar{\nu}_e$	$2 \times e^+e^-$ decay
$\to \nu_\mu \bar{\nu}_\mu$	$2 \times \mu^+\mu^-$ decay
$\to \nu_\tau \bar{\nu}_\tau$	$2 \times \tau\bar{\tau}$ decay
$\to \nu_{new} \bar{\nu}_{new}$	$N_{\substack{new \\ neutrino}} \times 2 \times e^+e^-$ decay

Total branching fraction into neutrinos

$$= (2+2+2) \times e^+e^- \text{ fraction} + N_{\substack{new \\ neutrino}} \times 2 \times e^+e^- \text{ fraction}$$

$$\text{Ratio of } \left[\frac{Z \to \nu\bar{\nu}}{Z \to e^+e^-}\right] = \left[6 + 2N_{\substack{new \\ neutrino}}\right] \text{ decays}$$

Figure 4

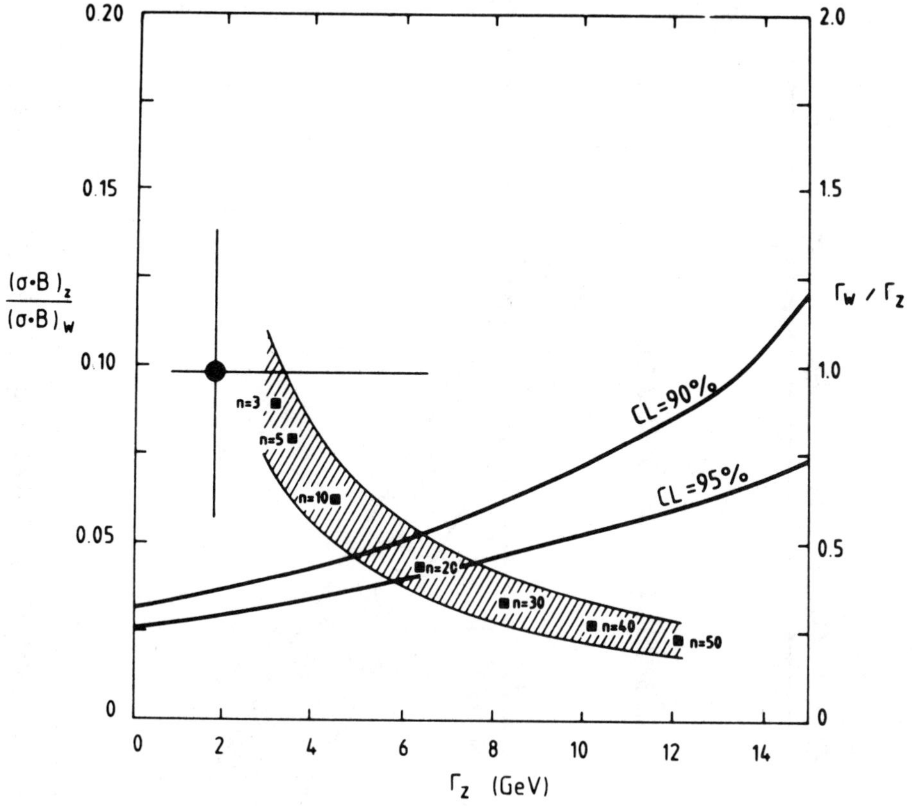
0.20
0.15
0.10
0.05
0
2.0
1.5
1.0
0.5
0
(σ•B)z / (σ•B)w
Γw / Γz
CL=90%
CL=95%
n=3
n=5
n=10
n=20
n=30
n=40
n=50
0
2
4
6
8
10
12
14
Γz (GeV)

Figure 5

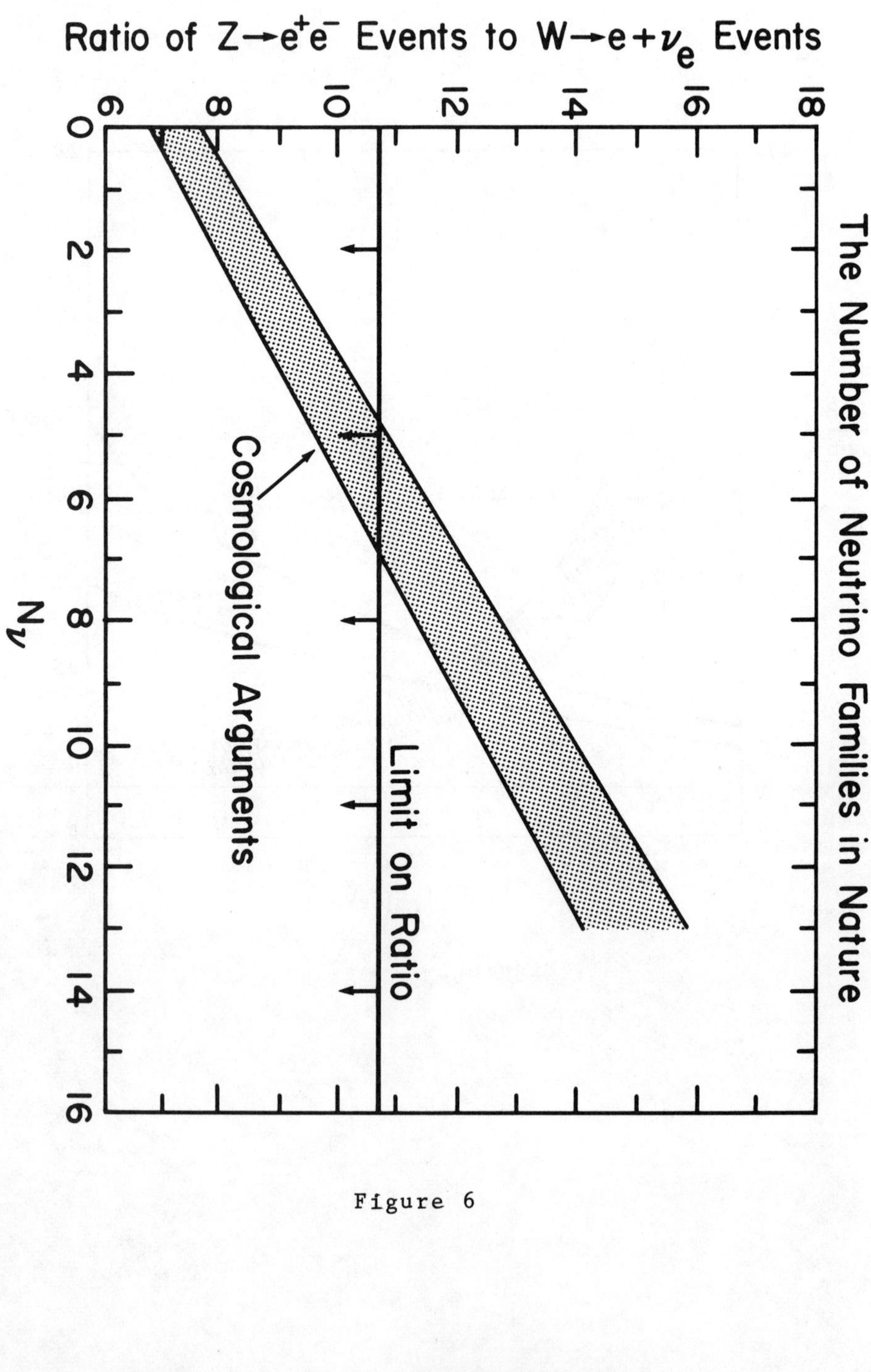
The Number of Neutrino Families in Nature
Ratio of Z→e⁺e⁻ Events to W→e+ν_e Events
Limit on Ratio
Cosmological Arguments
N_ν
0 2 4 6 8 10 12 14 16
6 8 10 12 14 16 18

Figure 6

standard model. The ratio of W and Z total widths can then be "measured" by forming the ratio

$$\frac{N_{Z \to \ell\ell}}{N_{W- \to \ell\nu_{\ell j}}} = \left(\frac{\Gamma_{Z \to \ell + \bar{\ell}}}{\Gamma_{W \to \ell + \nu_\ell}}\right) \bullet \left(\frac{\Gamma_W}{\Gamma_Z}\right) \bullet \frac{\sigma_Z}{\sigma_W}\frac{Q^2}{Q^2}$$

where $N_{Z \to \ell\bar{\ell}}$, $N_{W \to \ell\nu_\ell}$ refer to the corrected number of Z and W events into the partial channels $\ell\bar{\ell}$ and $\ell\nu_\ell$, respectively, and σ_Z, σ_W refer to the production cross sections and Q^2 is the momentum squared of the recoiling W and Z particles. Γ_W and Γ_Z are the total widths of the W and Z, respectively. The theoretical predictions for σ_Z and σ_W follow from the coupling of the W, Z to the $u\bar{u}$, $d\bar{d}$ and $u\bar{d}$ quark-antiquarks in the beam particles.

The calculation of $\Gamma_{W \propto \ell\nu}$ and $\Gamma_{Z \to \ell\bar{\ell}}$ are straight forward in the standard model.

$$\Gamma_{W \to \ell\nu_\ell} = \frac{G\, m_W^3}{6\sqrt{2}\pi}$$

$$\Gamma_{Z \to \ell\bar{\ell}} = \frac{G\, m_W^3}{12\sqrt{2}\pi}(1 - 4s^2 + 8s^4)$$

where as the calculations for Γ_W, Γ_Z total widths depends on the number of leptonic neutrinos flavors, t quark mass, the Weinberg angle $s^2 = \mathrm{Sin}^2\theta_w$ and the mass of the W and Z. We use M_W, M_Z and s^2 determined from UA1 experimental data. The ratio of Γ_W/Γ_Z can be reliably calculated on the basis of the assumed t quark mass and the number of neutrino flavors. A very precise measurements of Γ_W/Γ_Z and knowledge of the t quark mass allows for a determination of N_ν, the number of neutrino families in nature.

In the limit that $u_p = d_p = \bar{u}_{\bar{p}} >> s, \bar{s}$ the rate of cross sections becomes

$$\left[\frac{\sigma_Z}{s_{W^+} + \sigma_{W^-}}\right]_{\bar{p}p} = \left[\frac{1}{4} - \frac{2s^2}{3} + \frac{8s^4}{9}\right]\int_{\tau_Z}^{1} dx\, \frac{[u(x)\ \bar{u}(x)]}{x}$$

$$+\left[\frac{1}{4} - \frac{s^2}{3} + \frac{2}{9}s^4\right]\int_{\tau_Z}^{1} dx\frac{[d(x)\bar{d})x]}{x}$$

$$\int_{\tau_W}^{1} dx\left\{\frac{[u(x)\bar{d}(x)]}{x}\right\}\cos^2\theta_c + \left[\frac{u(x)\bar{s}(x)}{x}\right]\sin^2\theta_c$$

which gives

$$\left[\frac{1}{2} - s^2 + \frac{10}{9} s^4\right] \frac{1}{\cos^2 \theta_c} \left(\frac{\tau_Z}{\tau_W}\right) = 0.27$$

if we neglect the strange quark contribution and the different x dependence of u and d quarks. Thus it is plausible that the ratio of $[\sigma_Z/\sigma_{W^+} + \sigma_{W^-}]$ can be reliably calculated for $\bar{p}p$ collisions since the dominant contribution comes from the SU(2) x ν(1) parameters and the kinematic factors. We expect corrections to higher order QCD process to scale approximately like the first order calculations.

We conclude that for $\bar{p}p$ collisions the ratio

$$\left(\frac{\sigma_Z}{\sigma_{W^+} + \sigma_W}\right) [Q^2]$$

is likely reliably calculated. As $Q^2 \to 0$ the Drell-Yan $q\bar{q}$ annihilation graphs should give the dominate production cross section. Higher order QCD processes should give rise to a Q^2 distribution for the W, Z which has already been observed in the UA1 data.

We believe that the ratio $\sigma_Z/\sigma_{Z^+} + \sigma_{W^-}$ can be more reliably estimated than the individual cross sections since many of the uncertainties drop out of the ratio calculations; the ratio depends mainly on the semi-weak coupling of the $u\bar{u}$, $d\bar{d}$, and $u\bar{d}$ to the Z and W status.

In the standard model with a t-quark of mass of 40 GeV and 3 generations of neutrinos the ratio of W to Z width is exactly predicted to better than a few percent. If there are additional decays of the W or Z into new types of particles, such as supersymmetric particles, the ratio of widths will probably depart from these predictions. In Fig. 4 we report a result of Barger et al. that summarizes the present limits on the ratio of Γ_W/Γ_Z, and the theoretical uncertainties due to the measurement errors on the SU(2) x U(1) parameters. If we take the viewpoint that the theoretical limits on N_ν from big bang cosmology are correct the ratio of Γ_W/Γ_Z is rather precisely predicted.

The primordial abundance of ^{4}He is very sensitive to the number of neutrino flavors since in the early universe additional species of light neutrinos result in a faster universal expansion rate leaving more neutrnos to be bound in ^{4}He. The number of species inferred from the primardrel ^{4}He abundance also depends weakly on the exact neutron half life and the total density of baryons in the universe. Yang et al., have shown that the number of families is less than or equal to 4 (see ref. 4). Their calculations strongly disfavor even 5 families, a remarkable precise deduction for a cosmologically based calculation. Present collider data are in agreement with this result so far.

References (3)

1. ARNISON, G., et al. (1983). Phys. Lett. 122B, 103.
   ARNISON, G., et al. (1983). Phys. Lett. 126B, 398.

2. BANNER, M., et al. (1983). Phys. Lett. 122B, 476.
   BAGNANA, P., et al. (1983). Phys. Lett. 129B, 130.

3. ARNISON, G., et al. (1984). Phys. Lett. 147B, 222.

4. See for example, M. Turner, in Proc. Telemark Neutrino Mass Miniconference, ed. by V. Barger and D. Cline, P. 73, University of Wisconsin Report (1980).

5. CLINE, D. and RUBBIA, C. (1983). Phys. Lett. (1983) and CERN EP/83-61.

6. UA1 improvement program proposal SPSC/83-48/84-72.

7. The 1984 run provided a net integrated luminosity of 270 nb^{-1}.

8. See the UA2 report from this meeting.

9. Private communication, A. Sirlin and W. Marciano (1985).

10. This method was proposed by F. Halzen and K. Mursula, Phys. Rev. Lett. 51, 857 (1983); D. Cline and J. Rohlf, CERN report (unpublished); K. Hikasa, Phys. Rev. D 29, 1939 (1984).

11. BARGER, V. and HALZEN, F. (1985). University of Wisconsin Preprint.

12. CLINE, D. and MOHAMMADI, M. UA1 TN 1984.

13. ARNISON, G., et al. (1984). Phys. Lett. 134B, 469 (1984).

14. ARNISON, G., et al. (1985). Submitted to Phys. Lett.

4. The Search for New Quarks-Evidence for the t Quark from $W \to t\bar{b}$

The large center of mass energy available and the possibility of studying W and Z decays provide a unique chance to observe heavy quarks--either the t quark to complete the three generations of quarks or quarks from the 4th generation. The t quark is searched for in the decay $W \to t\bar{b}$ and some evidence now exists from the UA1 experiment.

The 4th generation quarks might be detected through the decay of the W as will be

$$W \to c + b'$$

However such processes are forbidden in the standard model. The other process to produce the 4th generation particles is

$$q + \bar{q} \to b' + \bar{b}', \text{ quark fusion}$$

In this chapter we give details of the search for the t quark in the UA1 experiment.

The present chapter deals with the search for the sixth quark, the t-quark (top or truth), which completes the family of the weak currents with a third doublet ($t\bar{b}_c$) which manifests itself in the decay $W^+ \rightarrow t\bar{b}_c$ (and also $W^- \rightarrow \bar{t}b_c$), provided $m_t < m_W - m_b$. Previous, unsuccessful searches for the t-quark in e^+e^- colliding beams [1] have established a mass limit $m_t > 22$ GeV/c^2. Using the available energy in the decay $W^\pm \rightarrow tb_c$, we can extend the search up to masses of about 65 GeV/c^2, since the t-quark is produced in association with a relatively light quark. For $m_t = 40$ GeV/c^2 the reduction factor with respect to massless quarks is 0.71, giving, before detection cuts, an expectation of (181 ± 20) $W^\pm \rightarrow tb_c$ decays from the integrated luminosity in the 1983 run.

We concentrate on the semileptonic decay channels of the t-quark:

$$W^+ \rightarrow t\bar{b}_c(t \rightarrow \ell^+\nu b_c) \quad \ell \equiv (\text{electron,muon})$$

(and the corresponding reaction for W^-). In spite of the smaller number of events due to the additional semileptonic branching ratio, this channel is chosen for the clean signature it provides, i.e. two jets, a lepton, and some missing transverse energy (ν) [2]. These events have several features which permit them to be identified and separated from other sources of background:

(i) The invariant mass of the ($b_c\bar{b}\ell\nu$) system must peak around the W mass. Replacing the neutrino momentum p_ν by the measured missing transverse energy ΔE_m [5] [≠2] broadens the peak somewhat, with a small shift of the average mass value.

(ii) The invariant mass of the ($b_c\ell\nu$) system must be compatiable with a common value, namely m_t [≠3] (to a good approximation after the substitution $p_\nu \rightarrow \Delta E_m$).

(iii) The t-quark is heavy ($m_t > 22$ GeV/c^2) and therefore relatively slow in the laboratory. Angles between particles from the decay in the laboratory are very wide. Furthermore, the lepton momentum has a large component normal to be b_c jet, p_n. In the case of decays of lighter quarks (b,c), $p_n \leqslant (m_b/2, m_c/2)$.

(iv) The recoiling b_c jet has a characteristic jacobian peak in the transverse energy distribution, which makes it possible, in most cases, to distignish it from the other, lower-energy b_c jet.

As discussed in detail further on, a simple set of topological cuts on the event configuration enables us to extract an essentially background-free event sample. However, the number of surviving events is also considerably reduced. However, the number of surviving events is also considerably reduced. For a semileptonic branching ratio ~1/9 ("naive" prediction based on lepton and quark counting) and $m_t = 40$ GeV/c^2, we expect (20±2.2) electrons (both signs) and an equal number of muons. With reasonable cuts on the transverse energy of jets and leptons [$E_T(\bar{b}_c) > 8$ GeV; E_T $(b_c) > 7$ GeV; $p_T(\ell) > 12$ GeV/c] we

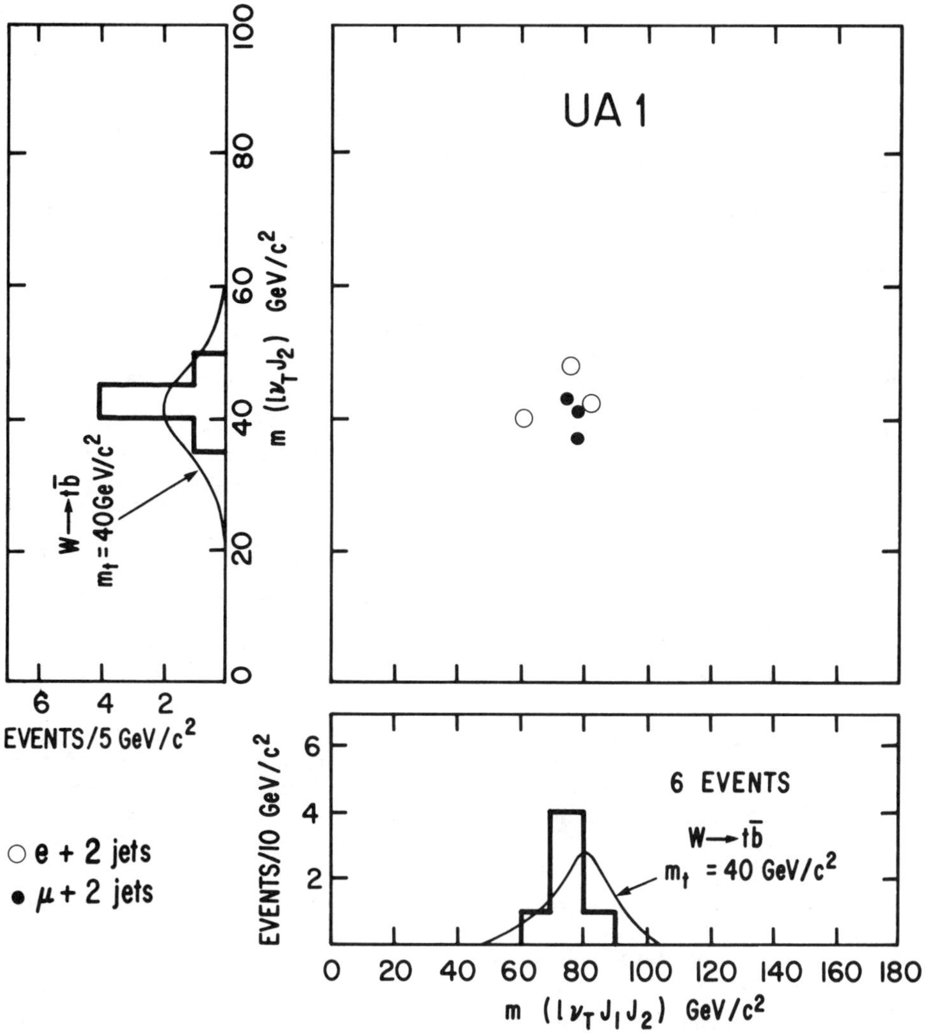
UA 1
m $(l\nu_T J_2)$ GeV/c²
W → tb̄
m_t = 40 GeV/c²
EVENTS/5 GeV/c²
○ e + 2 jets
● μ + 2 jets
6 EVENTS
EVENTS/10 GeV/c²
m $(l\nu_T J_1 J_2)$ GeV/c²

Figure 7

arrive at (4 ± 0.3) events for each leptonic channel, before geometrical and track isolation cuts.

Searching for this small number of events deserves some remarks:

(i) Both the muon and the electron samples were used in order to increase the significance of the result. Evidence of an effect must rely on its independent observation in both decay channels.

(ii) The electron and muon identification were considerably improved with respect to the previous search for leptonic IVB decays since now the signal is only ~1/20 of the $W \to \ell + \nu$ rate and the average lepton energy is a factor of ~3 smaller, which greatly enhances the probability of hadrons simulating the leptonic signature. Furthermore, the event topology requires a dominant jet activity, thus enhancing QCD associated backgrounds.

Finally, a residual leptonic signal due to $W \to t\bar{b}_c$ decays must be clearly separated from production of ($b\bar{b}$) and ($c\bar{c}$) with subsequent associated semileptonic decays, which are copious sources of leptons and jets. The cross section and kinematics of heavy quark pair-production via gluon-related and quark-related strong interaction graphs are relatively poorly known at collider energies. In order to extract from the data the information needed to reliably evaluate the expected background from this source, a parallel analysis has been performed aimed at strong interaction production of heavy quarks.

We now proceed to examine the physical origin of the six events with a lepton (muon or electron) and two jets. In Fig. (7) we show the effective mass of the lepton, two jets, and the transverse component of the neutrino. A very sharp peak can be observed at a value corresponding to the W mass, once we have allowed for the small shift and broadening arising from the neglect of the longitudinal component of the neutrino momentum.

The decay hypothesis $W \to tb$ with $m_t \simeq 40$ GeV/c^2 describes all kinematical distributions [3] very well. The rate of occurrence of the events, the number of $W \to e\nu$ decays, and the Monte Carlo determined detection efficiency can be combined in order to evaluate the top semileptonic branching ratio, which is 0.23 ± 0.09, to be compared with 0.13 for b-quarks and 0.12 for the lighter, charmed quarks.

Finally, we quote a systematic error of ±10 GeV/c^2 in the mass evaluation. This is primarily due to uncertainties in the reconstruction of jets and the determination of the associated parton four-vectors.

The UA1 group has observed a clear signal in the channel of an isolated large-transverse-energy lepton plus two or three associated jets. The two-jet signal has an over-all invariant mass clustering around the W mass, indicating a novel decay of the W. The rates and features of the two-jet events do not satisfy the expectations for charm and beauty decay. They are, however, consistent with the process $W \to t\bar{b}$, where t is the sixth "top" quark of the Cabibbo current. If this is indeed the case, then the mass of the top is bounded between 30 and 50 GeV/c^2. We stress that the present uncertainty in the ($\ell\nu_T j_2$) mass is due to the determination of the jet energies, and that more statistics are needed to confirm this conclusions and the true nature of the effect observed.

References (4)

1. TASSO Collab., M. Althoff et al., Phys. Lett. 138B (1984) 441; S. Yamada, Proc. 1983 Intern. Symp. on Lepton and photon interactions at high energies (Cornell, 1983); and DESY report 83-100 (1983);
S. L. Wu, Phys. Rep. 107 (1984) 59).
2. UA1 Collab., G. Arnison et al., Phys. Lett. 139B (1984) 115.
3. V. Barger et al., CERN preprint Ref. TH. 3972 (1984) and published in Phys. Lett.

5. The Search for New Leptons: $W \to L + \nu_L$

Recently the author and C. Rubbia proposed to search for massive leptons through the W decay.[1] Fig.(8) shows the mass range for which this search can be sensitive.

The decay width of the W into the $L + \nu_L$ channel is given by

$$G(W \to L+\nu_L) = (GM_W^3/6\sqrt{2}\pi)[1-(M_L^2/M_W^2)^2][1+1/2(M_L^2/M_W^2)].$$

This width is plotted in Fig. (8). Note that L leptons of mass up to 60 GeV/c^2 give an appreciable partial decay width. (The decays of Z^o are less useful, reaching L mass values of ~25 GeV/c^2.)

We now turn to the properties of the decay leptons from the process

$$p\bar{p} \to W^\pm + X(W^\pm \to L^\pm + \nu_L)(L^\pm \to \ell^\pm + \nu_\ell + \nu_L).$$

For simplicity we consider leptons with a mass near $M_W/2$ in order to carry out simple analytic calculations (M_L = 40 GeV/c^2). However, we expect that extension to masses of ~20 GeV/c^2 and 60 GeV/c^2 will show the same characteristics. The $L^\pm$ produced in the decay of the $W^\pm$ are expected to be fully polarized, and thus we consider the decay of a polarized lepton,

$$L^+ \to \ell^\pm + \nu_\ell + \bar{\nu}_L,$$

via the standard (V-A) interaction. The angular distribution for the decay $\ell^\pm$ is given by

$$dW/d(\cos\theta)d\varepsilon = G^2M_L^5/192\pi^3[(1 - 3\varepsilon) + \cos\theta(1 - 3\varepsilon)]^2,$$

where $\varepsilon = p_\ell/(M_L/2)$ and θ is the angle between the L spin direction and

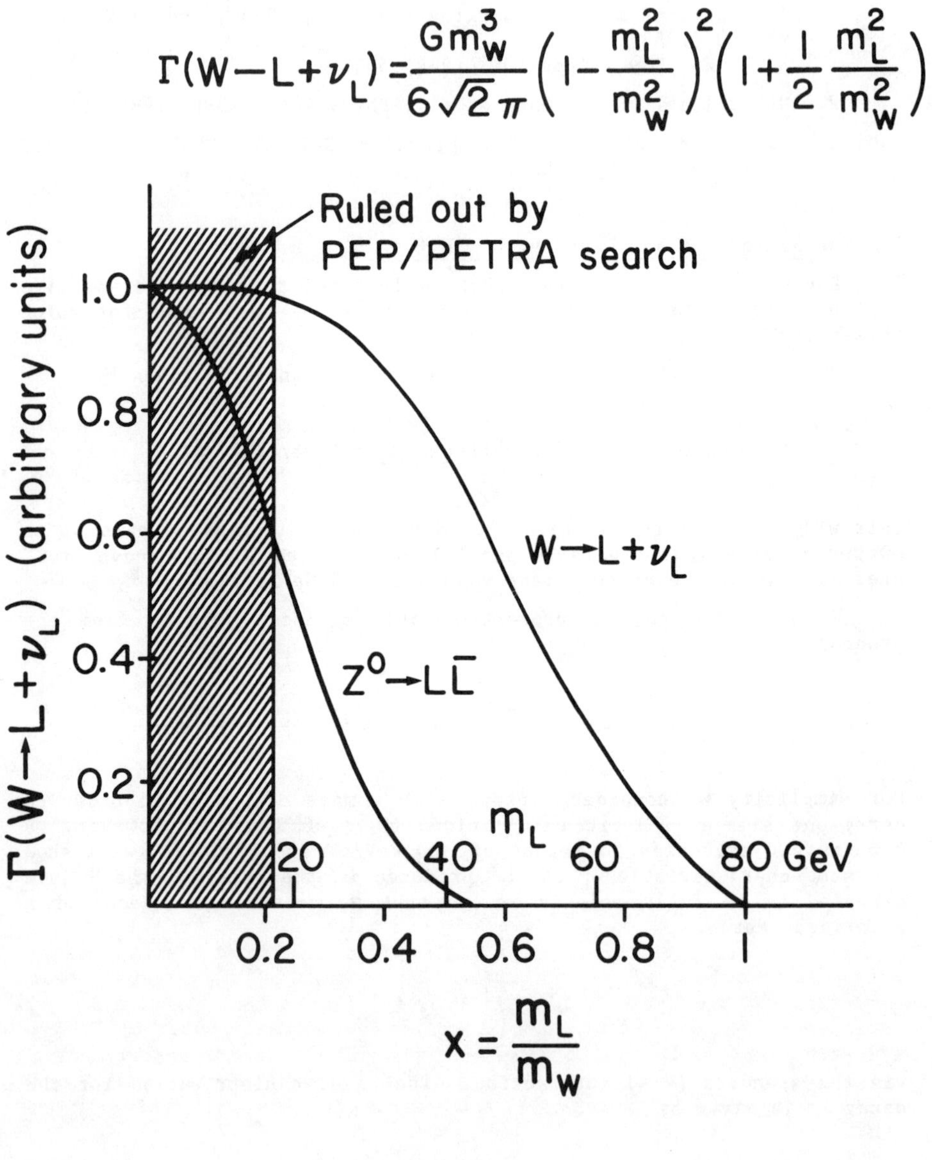
$\Gamma(W \to L + \nu_L) = \frac{G m_W^3}{6\sqrt{2}\pi}\left(1 - \frac{m_L^2}{m_W^2}\right)^2\left(1 + \frac{1}{2}\frac{m_L^2}{m_W^2}\right)$
Ruled out by
PEP/PETRA search
$\Gamma(W \to L + \nu_L)$ (arbitrary units)
1.0
0.8
0.6
0.4
0.2
$W \to L + \nu_L$
$Z^0 \to L\bar{L}$
m_L
20
40
60
80 GeV
0.2
0.4
0.6
0.8
1
$x = \frac{m_L}{m_W}$

Figure 8

$\ell^{\pm}$ momentum vector, and G is the Fermi constant. Integrating over ε, we find

$$dW/d(\cos\theta) = G^2M_L^5/192\pi^3[1/2 - 1/6\ \cos\theta].$$

Integrating over θ we find the standard distribution for ε; i.e.

$$dW = G^2M_L^5/92\pi^3[3 - 2\varepsilon]\varepsilon^2 d\varepsilon.$$

We take for the branching ratios of high-mass leptons the results of reference [2]. Note that the decay charged lepton from the L will be emitted in the backward direction compared to the L direction. This will modify the charge asymmetry of the charged leptons from $W^{\pm}$ decay and serve as a signature for the sequential lepton L.

There are several potential backgrounds for this observation. The main one comes from $W^{\pm}$ production with an unfavoured helicity due to sea ($q\bar{q}$ interactions. To illustrate this, we show as an example the expected $\ell^{\pm}$ angular distribution for the FNAL $p\bar{p}$ collider energy where the sea contribution is expected to be much larger. Thus the sea-quark distribution can produce a backward peaking in the angle-symmetry plot. However, the shape and magnitude of this contribution will not generally fit the $L^{\pm}$ production and decay hypothesis. This illustrates the need to measure the charged lepton angular distribution at a variety of angles and different centre-of-mass energy, since the sea contribution is expected to depend on c.m. energy in a chararteristic manner. We expect that at $p_T < 30$ GeV/c the L production and decay will change the asymmetry near $\cos\theta \sim 1$ for the FNAL $p\bar{p}$ collider energy as well.

It should also be possible to search for the L lepton through hadronic decays of the lepton, i.e.

$$L \rightarrow \nu_L + \text{(hadronic jet)}.$$

In this case events with a single high-p_T jet and large missing energy would be observed. However, there are potential backgrounds from heavy-flavour production and calorimetry fluctuation that must be addressed. Of course, the observation of the correct number of "wrong asymmetry" leptons in the intermediate p_T, as well as the corresponding number of jets plus missing energy events, would provide additional evidence for the $L^{\pm}$ hypothesis. A search is presently underway at CERN.

References (5)

1. D. Cline and C. Rubbia, Phys. Lett. 127B, 277 (1983).

2. Yung-Le Tsai, Phys. Rev. D4 (1971) 2821.

6. Multimuon Final States and $B\bar{B}$ Production

At the collider energy we expect copious production of $c\bar{c}$, $b\bar{b}$ and possibly tt pairs through QCD processes. As we will indicate later the measured (inferred) cross section for $b\bar{b}$ production is greater than 2 μb at 540 GeV. The basic technique to detect the heavy quarks is through the multilepton final states

$$\bar{p}p \to b + \bar{b} + \ldots$$
$$\bar{b} \hookrightarrow \mu^+ \ldots \qquad b \hookrightarrow \mu^- \ldots$$

$$\bar{p}p \to b + \bar{b}$$
$$\bar{b} \hookrightarrow \bar{c} + \ldots, \quad \bar{c} \hookrightarrow \mu^- \ldots$$
$$b \hookrightarrow \mu^- \ldots$$

In contrast with W and Z production cross-sections, which can be calculated reliably and have been measured in the leptonic decay channels at the CERN $p\bar{p}$ Collider[1,4,5], the cross-sections for strong QQ production are subject to various uncertainties. However, once a threshold of $p_\tau > 3$ GeV/c is applied to the leptons, different predictions[3] agree that the dominant contribution to dileptons is expected to come from $b\bar{b}$ final states rather than $c\bar{c}$. This is due to the hard fragmentation function ($z \sim 0.8$) for b quarks. Whereas process[5] gives like-sign leptons from first-generation decays, processes[3] and [6] yield unlike-sign leptons. However, like-sign dileptons can also originate from process[3] either through second-generation decays ($b \to c \to \mu$) or B^o-$\bar{B}^o$ mixing. Our recent report of a large D* content in jets[6] suggests that $c\bar{c}$ pair production within the same gluon jet is significant[7]. This mechanism can produce like- and unlike-sign dimuons.

The UA1 apparatus has already been described elsewhere[8]. The components relevant to the identification and measurement of muons were discussed in detail in ref. 5. In brief, a muon emerging from the $p\bar{p}$ interaction region traverses the Central Detector (CD), a tracking chamber that determines the muon momentum by its deflection in the central dipole field of 0.7 T. After leaving the CD, the muon must penetrate the electromagnetic calorimeter, the magnetized hadron calorimeter, and additional iron shielding (typically a total of more than 9 absorption lengths), before reaching the muon chambers. The momentum error due to the measurement errors in the CD is typically $\Delta p/p \sim 0.5\% \times p$ (p in GeV/c). The muon momentum measurements in the final event sample are such that $1/p > 3.5\sigma\,(1/p)$, and thus the charges are unambiguously determined.

During the 1983 data-taking period about 2.5×10^6 events were recorded on tape for an integrated luminosity of 108 nb^{-1}. Of these, about 1.0×10^6 events were muon triggers. All the events were passed through a fast filter program, wich selected 7.2×10^4 muon candidates with $p_\tau > 3$ GeV/c. These events were reconstructed by the standard UA1 programs. An automatic selection program, which applied stricter CD track quality and CD/muon chamber matching requirements was used to select dimuon events. Applying the muon p_τ cuts (1), 49 candidate dimuon events were selected, and were scanned on a computer graphics display facility. Of these events, 8 were identified as cosmic rays, 4 as 'kins' (π, K decay) or poor matching between the CD and the muon chamber tracks, and 11 as leakage of hadron showers through cracks in the calorimeters, leaving 26 candidates after scanning.

Five events are Z^0 candidates and have been described in ref. [1]. The muon p_T of the other 21 events are shown in fig. 1a, along with the p_τ cuts applied in the selection.

We first describe the general features of the 21 events which survive the selection criteria. There are 17 unlike-sign muon pairs and 4 like-sign pairs. The transverse momenta of the individual muons range from the 3 GeV/c cut up to 15 GeV/c (fig. 9), with a distribution much flatter than expected from background such as $\pi, K \to \mu$ decays. It is remarkable that the event with the highest transverse muon momentum ($p_\tau > 10$ GeV/c for both muons) has a like-sign pair.

From the bias due to the muon p_T cuts (1) we expect high-mass pairs (masses above ~ 10 GeV/c^2), with the two muons opposite in azimuth, or low-mass pairs with considerable dimuon transverse momenta. This behaviour is illustrated in fig. 10, where we show the correlation between the mass and the p_τ of the dimuon system. In most of the events the muons are approximately back-to-back, and in only six events are both muons in the same hemisphere. However, events in the region roughly bounded by

$$[(p_\perp)^2 + (m^{\mu\mu})]^{1/2} \leqslant 10 \text{ GeV} \tag{7}$$

are heavily suppressed by the muon p_T cuts (1).

The isolation of the muons is an important tool for distinguishing between the Drell-Yan process and heavy-flavour decays. Muons arising from a process such as the Drell-Yan one should not be embedded in significant hadronic activity, in constrast with c or b quark decays. The overall activity observed in the dimuon events cover a very wide range of transverse energies. Some events have several accompanying jets in the central region, with transverse energies. Some events have several accompanying jets in the central region, with transverse energies up to 60 GeV/c; others are completely quiet. We classify the muon isolation by calculating the ΣE_τ around the muon, i.e. the sum of the transverse energy of all the calorimeter cells within a cone of $\Delta R < 0.7$ about the muon in pseudorapidity (η)-azimuth (ϕ) space, where

$$\Delta R = (\Delta\eta^2 + \Delta\phi^2)^{1/2}. \tag{8}$$

The average energy deposition due to the muon itself is subtracted from the sum (11). The sum ΣE_τ for the muons in each event is plotted in fig. 11, which shows that there is a class of events where both muons

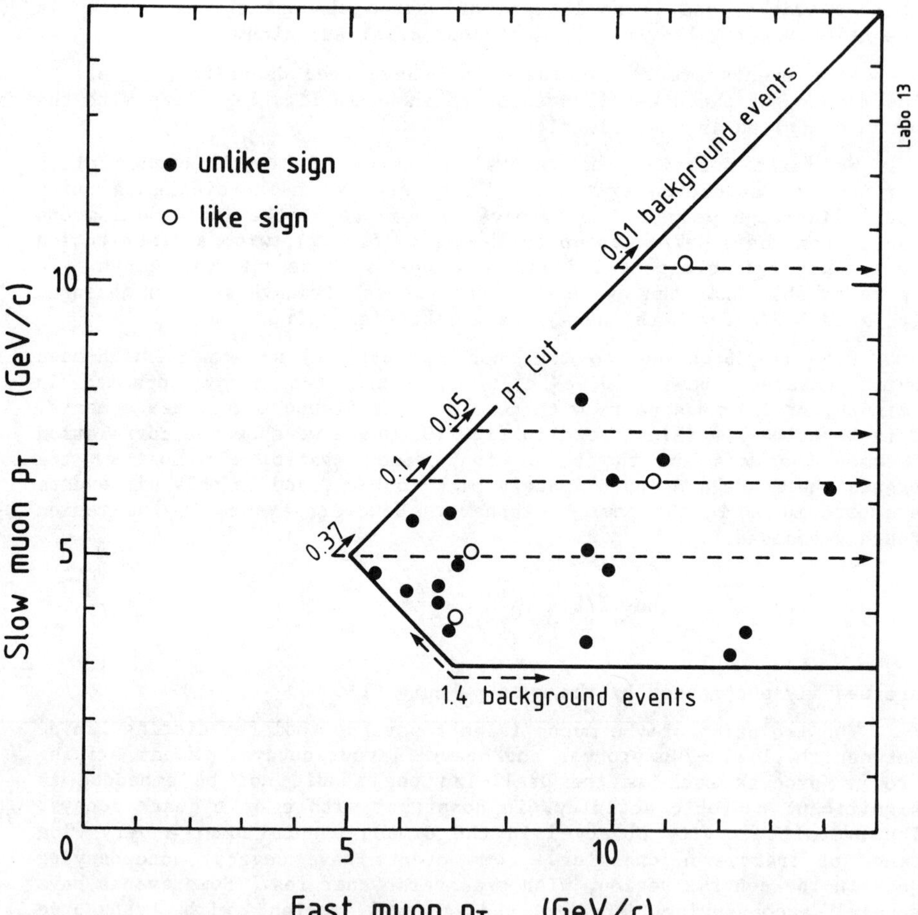
unlike sign
like sign
0.01 background events
p_T Cut
0.05
0.1
0.37
1.4 background events
Slow muon p_T (GeV/c)
Fast muon p_T (GeV/c)
0
5
10
Labo 13

Figure 9

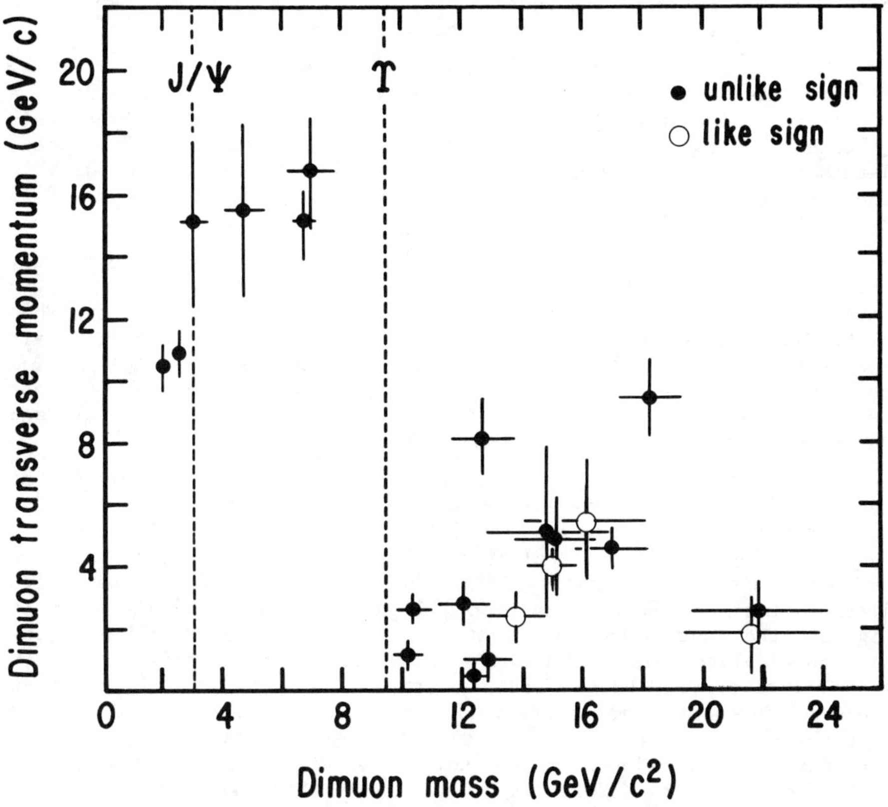

J/Ψ
Υ
unlike sign
like sign
Dimuon transverse momentum (GeV/c)
Dimuon mass (GeV/c²)
0
4
8
12
16
20
24

are clearly in jets. Taking an isolated muon to be one with $\Sigma E_\tau < 4$ GeV, we find 9 (3) unlike- (like-) sign events with both muons isolated, 3 (0) events with one isolated muon, and 5 (1) events with neither muon isolated. The dotted curve in fig. 2b is deduced from minimum-bias events in which the ΔR cone was randomly distributed in pseudorapidity and azimuth, and the fluctuations in the energy loss of the muon [10] have been folded into the distribution. It indicates the level of activity present from the underlying event. The dashed curve is a Monte Carlo simulation for $b\bar{b}$ and $c\bar{c}$ production [12]. The heavy-flavour curve is normalized to 9.9 events (according to subsection 6.2), and the minimum-bias curve to 11.1 events, so that the sum of the two curves (solid line) correspond to 21 events.

Heavy-flavour decays are often accompanied by strange particles. We have searched for K^o_s, Λ's, and Λ's by identifying V^o's. A constrained fit was done on candidate V^o's for each hypothesis (including a γ fit for conversions). A χ^2 cut and a minimum decay length of 0.5 cm were required for consideration. The number of V^o's produced per minimum-bias event for this rapidity range is 0.15 ± 0.04 (0.45 ± 0.09) for Λ/Λ (K) [14].

Heavy-flavour production of dimuons will also yield two neutrinos, which in principle should give rise to missing energy. However, the neutrinos will generally have low p_τ because of the bias from the muon p_τ cuts. Furthermore, if the neutrinos are back-to-back their transverse momenta will tend to cancel out in the missing E_τ. The observed missing energy is, in general, small. Because of the poor resolution at small E_τ^{miss}, it is not surprising that the average deviation from zero is in fact only 1.4σ.

In another class of events the muons come from heavy-flavour decays and hence are expected to be accompanied by hadronic activity. If we consider the events with at least one non-isolated ($\Sigma E_\tau > 4$ GeV) muon as being a relatively clean sample of heavy-flavour events, then we have nine such events (fig. 11b). The calculated isolation distribution for $b\bar{b}/c\bar{c}$ production [12] is compared with the data in fig. 2b, the theoretical distribution being normalized relative to the nine non-isolated events. The two distributions are consistent. With this normalization, heavy flavours are estimated to contribute about one event where both muons are isolated. From the Monte Carlo we find that $b\bar{b}$ is the dominant component with our cuts ($c\bar{c}$ contributes ~ 10%), and the cross-section for producing $b\bar{b}$ with both quarks having $p_\tau > 5$ GeV/c and $|\eta| < 2$ is 2.1 ± 0.8 μb with the above normalization.

There is one like-sign event among the nine non-isolated events. Like-sign events can be produced in the cascade process from $b\bar{b}$, $t\bar{t}$, and $t\bar{b}$ final states, and also from the primary $b\bar{b}$ quarks with B^o–$\bar{B}^o$ mixing. Many events have additional high-p_τ strange particles, this favours the heavy-flavour interpretation.

Three events with like-sign dimuons are observed (Q,S,T) in which both muons are isolated according to our criterion $\Sigma E_\tau < 4.0$ GeV. In all these events the muon momenta are 6 standard deviations or more from infinite momenta; hence the charges are very well determined. Event T may come from a second-generation decay of a b quark, since one muon is with an identified CD jet. Also one muon p_τ is small (the muons are near the kinematical cuts), as would be expected to result from a second-generation decay of a b quark. Events Q and S have no jets, and the higher muon p_T's make second-generation $b\bar{b}$ decay unlikely. Both events have a similar topology. In event Q (S) the $\mu^+\mu^+(\mu^-\mu^-)$ are back-to-back, and a Λ (Λ) is found at about 90° to both muons in the transverse plane. In event Q, both muons have very high transverse momenta (larger than 10 GeV); as a consequence the background is small ($< 3 \times 10^{-3}$ events). Calculations have failed to reproduce events with characteristics of event Q from $b\bar{b}$ production with or without B^o-$\bar{B}^o$ mixing. We expect that requiring both muons to have $p_T > 9$ GeV/c will give about 0.2 events with unlike signs and about 0.03 with like signs. We estimate of the order of 10^{-4} events with two isolated like-sign muons. The corresponding rates are about a factor of 10 larger for event S. By invoking mixing, the like-sign can be increased by about a factor of 2.

Fig. 11a

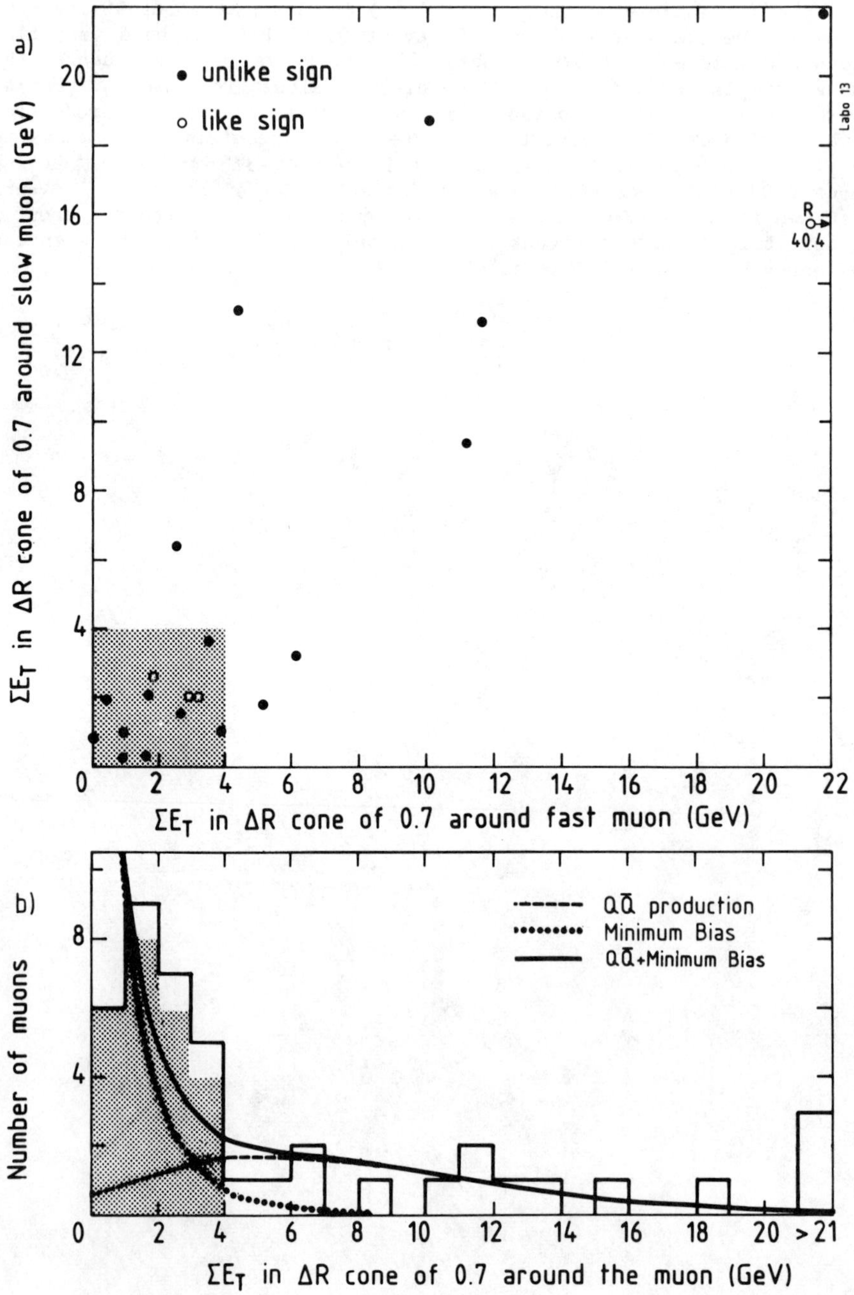

Fig. 11b

References (6)

1. G. Arnison et al. (UA1 Collaboration), Phys. Lett. 147B (1984) 241.

2. S. D. Drell and T. M. Yan, Phys. Rev. Lett. 25 (1970) 316.

3. See for example:

S. Pakvasa et al., Phys. Rev. D20 (1979) 2862.
F. E. Paige and S. D. Protopopescu, ISAJET program, BNL 29777 (1981).
B. L. Combridge, Nucl. Phys. B151 (1979) 429.
Z. Kunszt and E. Pietarinen, Nucl. Phys. B164 (1980) 45.

4. G. Arnison et al. (UA1 Collaboration), Phys. Lett. 126B (1983) 398.
G. Arnison et al. (UA1 Collaboration), Phys. Lett. 129B (1983) 273.
P. Bagnaia et al. (UA2 Collaboration), Phys. Lett. 129B (1983) 130.
P. Bagnaia et al. (UA2 Collaboration), Z. Phys. C24, (1984).
For a review on weak boson production, see E. Radermacher, preprint
CERN-EP/84-41 (1984), to be published in Progress in Particle and
Nuclear Physics.

5. G. Arnison et al. (UA1 Collaboration), Phys. Lett. 134B (1984) 469.

6. G. Arnison et al. (UA1 Collaboration), Phys. Lett. 147B (1984) 222.

7. F. Halzen and F. Herzog, Phys. Rev. D30 (1984) 2326.

8. M. Barranco Luque et al., Nucl. Instrum. Methods 176 (1980) 175.
M. Calvetti et al., Nucl. Instrum. Methods 176 (1980) 255.
M. Calvetti et al., IEEE Trans. Nucl. Sci. NS-30 (1983) 71.
M. Corden et al., Phys. Scr. 25 (1982) 5 and 11.
K. Eggert et al., Nucl. Instrum. Methods 176 (1980) 217.

9. G. Arnison et al. (UA1 Collaboration, Phys. Lett. 132B (1983) 214. The corrections to jet energies and errors are discussed in 'Correction tables and error estimates for jet energies and momenta', M. Della Negra et al., UA1 Technical Note UA1/TN/84-43 (1984) (unpublished).

10. A 10 GeV/c muon is expected to deposit about 1.7 GeV in the central calorimeters at normal incidence). This is an important effect when examining muon isolation or jets with muons. Consequently, the average expected energy deposition is substracted from the calorimeter cells traversed by the muon. However, for the central hadron calorimeters the energy loss for a 10 GeV/c muon is a Landau distribution peaked at 1.4 GeV and with a r.m.s. of 0.5 GeV (the electromagnetic calorimeters peak at 0.3 GeV, and hence have a small effect). We can then expect that about 2 out of 42 muons will deposit 2

GeV or more in excess of the expected value. This is clearly significant when discussing isolation.

11. F. E. Paige and S. D. Protopopescu, ISAJET program, BNL 31987 (1982).

12. These calculations were made using the Eurojet Monte Carlo program which contains all first-order (α_s^2) and second-order (α_s^3) QCD processes [see B. Van Eijk, UA1 Technical Note UA1/TN/84-93 (1984), unpublished, and A. Ali, E. Pietarinen and B. Van Eijk, to be published]. The heavy quark Q is fragmented using the parametrization of C. Peterson et al., Phys. Rev. D27 (1983) 105, and the V-A decay matrix elements.

13. G. Bauer et al., UA1 Technical Note UA1/TN/84-21 (1984) (unpublished).

14. K. Alpgard et al. (UA5 Collaboration), Phys. Lett. 115B (1982) 65.

15. G. Altarelli et al., preprint CERN-TH.4015/84 (1984).

Figure Captions

Fig. 1 a) The scatter plot of the p_T of the faster muon versus the p_T of the slower muon, with the kinematic boundary of the cuts (solid line). In addition, the number of expected background events integrated over the regions enclosed by the dashed line is indicated (see section 4).
b) The scatter plot of the dimuon mass versus the opening angle between the two muons in the transverse plane.
c) The scatter plot of the dimuon mass versus the dimuon transverse momentum. The positions of the J/ψ and the Υ resonances are indicated for reference.

Fig. 2: The scatter plot (a) shows the transverse energy in a ΔR cone of 0.7 around the fast muon versus the slow muon. The lower plot (b) is the transverse energy in a ΔR cone of 0.7 around each muon.[10] Curves show the distributions for the expectation from $b\bar{b}/c\bar{c}$ production (dashed line) [12], and a randomly orientated cone in minimum-bias events (dotted line). The sum of the two curves (solid line) corresponds to 42 muons (see text for normalizations).

Acknowledgements

I wish to thank members of the UA1 group for many discussions and to V. Barger, F. Halzen, A. Sirlin and W. Marciono for many discussions on theoretical matters relating to high energy $\bar{p}p$ collisions and W and Z bosons.

SUPERSYMMETRY, SUPERGRAVITY, AND PARTICLE PHYSICS*

Burt A. Ovrut†
Department of Physics, University of Pennsylvania
Philadelphia, Pennsylvania 19104

ABSTRACT

N = 1 global supersymmetry and supergravity are discussed using the superfield formalism. Supergravity particle physics theories are presented, emphasizing the role of gravitation in low energy physics.

Lecture 1: Supersymmetry

1) Supersymmetry Algebra

The 2-component Weyl spinors $\hat{Q}_\alpha$, $\hat{\bar{Q}}_{\dot{\alpha}}$ $(\alpha,\dot{\alpha}=1,2)$ and 4 momentum $\hat{P}_m$ (m=0,1,2,3) are generators satisfying the supersymmetry algebra:

$$\{\hat{Q}_\alpha,\hat{\bar{Q}}_{\dot{\alpha}}\} = 2\,\sigma^m{}_{\alpha\dot{\alpha}}\hat{P}_m$$

$$\{\hat{Q}_\alpha,\hat{Q}_\beta\} = \{\hat{\bar{Q}}_{\dot{\alpha}},\hat{\bar{Q}}_{\dot{\beta}}\} = 0 \qquad \sigma^o = \begin{pmatrix} 1 & 0 \\ 0 & 1 \end{pmatrix}$$

$$[\hat{P}_m,\hat{Q}_\alpha] = [\hat{P}_m,\hat{\bar{Q}}_{\dot{\alpha}}] = 0$$

$$[\hat{P}_m,\hat{P}_n] = 0 \tag{1.1}$$

2) Super Group

Define a super group element:

$$e^{i\{-x^m P_m+\theta^\alpha\hat{Q}_\alpha+\bar{\theta}_{\dot{\alpha}}\hat{\bar{Q}}^{\dot{\alpha}}\}} \equiv \sigma(x^m,\theta^\alpha,\bar{\theta}_{\dot{\alpha}}) \tag{1.2}$$

where x^m(m=0,1,2,3) is a space-time coordinate with the metric

$$\eta_{mn} = \begin{pmatrix} -1 & 0 & 0 & 0 \\ 0 & 1 & 0 & 0 \\ 0 & 0 & 1 & 0 \\ 0 & 0 & 0 & 1 \end{pmatrix}$$

and θ^α, $\bar{\theta}_{\dot{\alpha}}$ are anti-commuting parameters.

---

*Four lectures presented at the Mexican School of Physics, Ouaxtepec, Mexico December, 1985.

†On leave of absence from the Rockefeller University, New York, N.Y., 10021

0094-243X/86/1430097-25$3.00

We will use the notation:

$$\theta^{\alpha}\hat{Q}_{\alpha} = \theta\hat{Q} \quad ,$$

$$\bar{\theta}_{\dot{\alpha}}\hat{\bar{Q}}^{\dot{\alpha}} = \bar{\theta}\hat{\bar{Q}} \quad ,$$

Note that

$$\theta^{\alpha}Q_{\alpha} = \varepsilon_{\alpha\beta}\theta^{\alpha}Q^{\beta} = -\varepsilon_{\beta\alpha}\theta^{\alpha}Q^{\beta} = -\theta_{\alpha}Q^{\alpha} \quad .$$

Multiply two super group elements (<u>left</u> multiplication)

$$\sigma(0,\xi,\bar{\xi})\sigma(x,\theta,\bar{\theta}) = e^{i\{\xi\hat{Q}+\bar{\xi}\hat{\bar{Q}}\}}e^{i\{-xP+\theta\hat{Q}+\bar{\theta}\hat{\bar{Q}}\}} \quad .$$

Using the Campbell-Baker-Hausdorf formula

$$e^{A}e^{B} = c^{A+B+\frac{1}{2}[A,B]+\ldots.}$$

it is easy to show that

$$\sigma(\sigma,\xi,\bar{\xi})\sigma(x,\theta,\bar{\theta}) =$$

$$= e^{i\{-x\hat{P}+i(\xi\sigma\bar{\theta}-\theta\sigma\bar{\xi})\hat{P}+(\theta+\xi)\hat{Q}+(\bar{\theta}+\bar{\xi})\hat{\bar{Q}}\}}$$

$$= \sigma(x+i(\theta\sigma\bar{\xi}-\xi\sigma\bar{\theta}),\theta+\xi,\bar{\theta}+\bar{\xi}) \qquad (1.3)$$

Thus, the action of an infinitesimal supertransformation on Superspace is

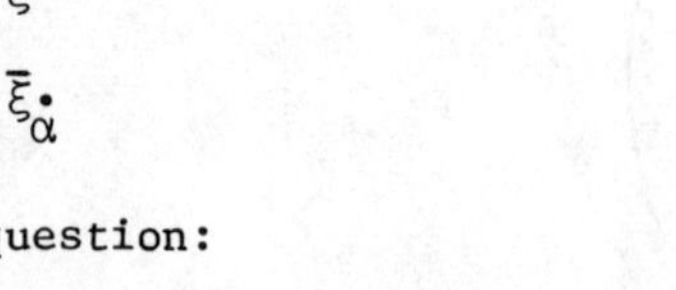

$$x^{m} \to x^{m} + i(\theta\sigma^{m}\bar{\xi}-\bar{\xi}\sigma^{m}\bar{\theta})$$

$$\theta^{\alpha} \to \theta^{\alpha} + \xi^{\alpha}$$

$$\bar{\theta}_{\dot{\alpha}} \to \bar{\theta}_{\dot{\alpha}} + \bar{\xi}_{\dot{\alpha}} \qquad (1.4)$$

Then we raise a question:

"Can we find explicit generators Q_{α}, $\bar{Q}_{\dot{\alpha}}$ and P_{m} (no hat!!) such that

$$e^{\xi Q+\bar{\xi}\bar{Q}}\, x^{m} = x^{m} + i(\theta\sigma^{m}\bar{\xi}-\xi\sigma^{m}\bar{\theta})$$

$$e^{\xi Q+\bar{\xi}\bar{Q}}\, \theta^{\alpha} = \theta^{\alpha} + \xi^{\alpha}$$

$$e^{\xi Q+\bar{\xi}\bar{Q}}\, \bar{\theta}_{\dot{\alpha}} = \bar{\theta}_{\dot{\alpha}} + \bar{\xi}_{\dot{\alpha}} \qquad ?"$$

The answer is Yes. They are:

$$Q_\alpha = \frac{\partial}{\partial\theta^\alpha} - i\,\sigma^m_{\alpha\dot\alpha}\bar\theta^{\dot\alpha}\partial_m$$

$$\bar Q^{\dot\alpha} = \frac{\partial}{\partial\bar\theta_{\dot\alpha}} + i\,\theta^\alpha\sigma^m_\alpha{}^{\dot\alpha}\partial_m \tag{1.5}$$

$$P_m = i\,\partial_m \qquad \text{(\underline{Left} Multiplication)}$$

These generators satisfy

$$\{Q_\alpha, \bar Q_{\dot\alpha}\} = 2\,\sigma^m_{\alpha\dot\alpha}(i\partial_m)$$

and all other (anti) commutation relations.

We can do the same thing for the right multiplication and obtain

$$D_\alpha = \frac{\partial}{\partial\theta^\alpha} + i\,\alpha^m_{\alpha\dot\alpha}\bar\theta^{\dot\alpha}\partial_m$$

$$\bar D_{\dot\alpha} = -\frac{\partial}{\partial\bar\theta^{\dot\alpha}} - i\,\theta^\alpha\sigma^m_{\alpha\dot\alpha}\partial_m \tag{1.6}$$

$$P_m = -\,i\partial_m \qquad \text{(\underline{Right} Multiplication)}$$

These generators satisfy all (anti) commutation relations.

3) Superfields

A superfield is defined as a mapping from superspace to the complex numbers (modulo some mathematical subtleties).

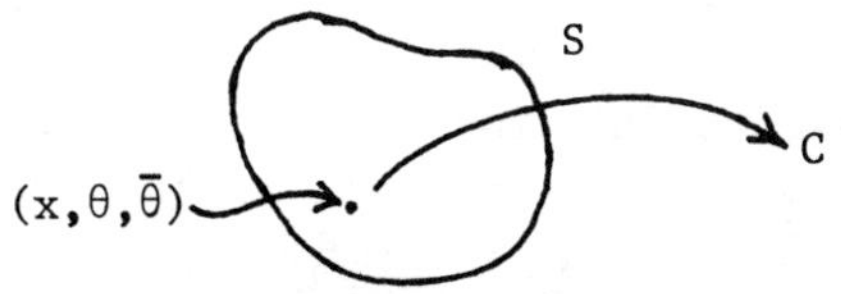

The Taylor expansion of $S(x,\theta,\bar\theta)$ in θ and $\bar\theta$ must terminate after a small, finite number of terms since $(\theta^1)^2 = -(\theta^1)^2 = 0$ etc. The most general form for S is:

$$S(x,\theta,\bar\theta) = f(x) + \theta\phi(x) + \bar\theta\bar\chi(x) + \theta^2 m(x)$$

$$+ \bar\theta^2 n(x) + \theta\sigma^m\bar\theta V_m(x) + \theta^2\bar\theta\bar\lambda(x) + \bar\theta^2\theta\psi(x)$$

$$+ \theta^2\bar\theta^2 d(x) \tag{1.7}$$

We call

$$f(x),\ \phi(x),\ \bar{\chi}(x),\ m(x),\ n(x),\ V_m(x),\ \bar{\lambda}(x),\ \psi(x) \text{ and } d(x)$$

<u>the component fields of superfield S.</u>

Note that

1) The sum of superfields is again a superfield,
2) The product of superfields is again a superfield.

Under a supersymmetric transformation

$$\delta_\xi S(x,\theta,\bar{\theta}) = \delta_\xi f(x) + \theta\delta_\xi\phi(x) + \ldots + \theta^2\bar{\theta}^2\delta_\xi d(x)$$
$$= (\xi Q+\bar{\xi}\bar{Q})S(x,\theta,\bar{\theta}) \quad .$$

Using equations (1.5) and (1.7), we can read off the transformation law for the component fields. However, in general these transform as a reducible multiplet (but not completely reducible).

Question: "Can we find superfields whose component fields transform irreducibly under supertransformations?"

Answer: "Yes, But we need constraints."

Chiral Superfields

If we demand

$$\bar{D}_{\dot{\alpha}}S = 0 \quad , \tag{1.8}$$

the most general form of the superfields of this type becomes

$$\phi(x,\theta,\bar{\theta}) = A(x) + \sqrt{2}\ \theta\psi(x) + \theta^2F(x) + i\ \theta\sigma^m\bar{\theta}\partial_m A$$
$$- \frac{i}{\sqrt{2}}\ \theta^2\partial_n\psi\sigma^n\bar{\theta} + \frac{1}{4}\ \theta^2\bar{\theta}^2\partial^2 A \quad . \tag{1.9}$$

This can be put in a simpler form in terms of

$$y = x + i\ \theta\sigma\bar{\theta} \tag{1.10}$$

$$\phi(y,\theta,\bar{\theta}) = A(y) + \sqrt{2}\ \ \theta\psi(y) + \theta^2F(y) \tag{1.11}$$

$\psi(x)$ and $A(x)$ are physical fields, while $F(x)$ is an auxiliary field.

Noting that Q_α and $\bar{Q}_{\dot{\alpha}}$ can be written in terms of y, the transformation property for the component fields of chiral supermultiplets can be read off from

$$\delta_\xi\phi(y,\theta,\bar{\theta}) = \delta_\xi A(y) + \theta\delta_\xi\psi(y) + \theta^2\delta_\xi F(y)$$
$$= (\xi Q+\bar{\xi}\bar{Q})\phi(y,\theta,\bar{\theta})$$

to be

$$\delta A(y) = \sqrt{2}\, \xi\psi(y)$$
$$\delta\psi(y) = \sqrt{2}\, \xi F(y) + \sqrt{2}\, i\sigma^n\bar{\xi}\partial_{yn}A(y)$$
$$\delta F(y) = -\, i\, \sqrt{2}\, \partial_m\psi\sigma^m\bar{\xi} \qquad (1.12)$$

These are the usual component field supersymmetry transformations. Note that these transformations are irreducible. If we also demand

$$D_\alpha\phi = 0$$

on a chiral superfield ϕ, then ϕ = const.

Vector Superfields

Another important class of irreducible superfields are vector superfields.

If we demand:

$$S = S^+ \quad , \qquad (1.13)$$

the most general form of superfields of this type becomes:

$$V(x,\theta,\bar{\theta}) = C + i\theta\chi - i\bar{\theta}\bar{\chi} + \frac{i}{2}\,\theta^2(M+iN)$$
$$-\,\frac{i}{2}\,\bar{\theta}^2(M-iN) - \theta\sigma^m\bar{\theta}V_m + i\theta^2\bar{\theta}(\bar{\lambda}+\frac{i}{2}\bar{\sigma}^m\partial_m\chi)$$
$$-\, i\,\bar{\theta}^2\theta(\lambda+\frac{i}{2}\sigma^m\partial_m\bar{\chi}) + \tfrac{1}{2}\,\theta^2\bar{\theta}^2(D+\tfrac{1}{2}\partial^2C) \qquad (1.14)$$

All component fields are functions of x. Fields C, M, N, V_m, D and C are real. As we have done for the chiral multiplets, we can find δC, $\delta\chi$, etc. and show that the component fields of V transform irreducibly under supertransformations.

Note that, for a chiral superfield ϕ, $\phi + \phi^+$ is not chiral but vector. Also

$$\phi + \phi^+ = \ldots - \theta\sigma^m\bar{\theta}\partial_m(-i)(A-A^+) + \ldots \quad .$$

Now consider

$$V + \phi + \phi^+ = \ldots - \theta\sigma^m\bar{\theta}\,(V_m - i\partial_m(A-A^+)) + \ldots$$

The r.h.s. shows the familiar transformation property for the U(1) gauge field. This leads us to define the superfield gauge transformation to be

$$V \rightarrow V + \phi + \phi^+ \qquad (1.15)$$

Under (1.15) the component fields of V transform as

$$C \to C + A + A^+$$

$$\chi \to \chi - i\sqrt{2}\,\psi$$

$$M+iN \to M + iN - 2\,iF$$

$$V_m \to V_m - i\partial_m(A-A^+)$$

$$\lambda \to \lambda$$

$$D \to D \tag{1.16}$$

The Wess-Zumino gauge is defined as that class of gauges for which

$$C = 0$$

$$\chi = 0$$

$$M+iN = 0$$

This choice of gauge breaks the supersymmetry since the supersymmetry transformation is irreducible. Note that the freedom $V_m \to V_m + \partial_m\Lambda$ is still allowed in this gauge. Thus in the Wess-Zumino gauge, V is given by

$$V = -\,\theta\sigma^m\bar{\theta}V_m + i\theta^2\bar{\theta}\bar{\lambda} - i\bar{\theta}^2\theta\lambda + \tfrac{1}{2}\theta^2\bar{\theta}^2 D \tag{1.17}$$

Here V_m and λ are a gauge boson and its corresponding gaugino, while D is an auxiliary field.

Before proceeding, we shall calculate some powers of V (for practice in the algebra of anticommuting variables).

$$V^2 = -\tfrac{1}{2}\,\theta^2\bar{\theta}^2 V_m V^m$$

$$V^3 = 0 \tag{1.18}$$

Proof:

$$V^2 = \theta^\alpha\sigma^m{}_{\alpha\dot{\alpha}}\bar{\theta}^{\dot{\alpha}}V_m\,\theta^\beta\sigma^n{}_{\beta\dot{\beta}}\bar{\theta}^{\dot{\beta}}V_n$$

$$= \theta^\alpha\bar{\theta}^{\dot{\alpha}}\theta^\beta\bar{\theta}^{\dot{\beta}}\sigma^m{}_{\alpha\dot{\alpha}}\sigma^n{}_{\beta\dot{\beta}}V_mV_n$$

$$= -(\theta^\alpha\theta^\beta)(\bar{\theta}^{\dot{\alpha}}\bar{\theta}^{\dot{\beta}})\sigma^m{}_{\alpha\dot{\alpha}}\sigma^n{}_{\beta\dot{\beta}}V_mV_n$$

Noting that $\theta^\alpha\theta^\beta$ is an antisymmetric tensor which must be proportional to $\varepsilon^{\alpha\beta}$,

$$\theta^\alpha\theta^\beta = C\varepsilon^{\alpha\beta}$$

$$\theta^2 = C\varepsilon^{\alpha\beta}\varepsilon_{\alpha\beta} = - C\varepsilon^{\alpha\beta}\varepsilon_{\beta\alpha} = - C\,\delta^\alpha{}_\alpha$$

$$= - 2C$$

$$C = - \tfrac{1}{2}\,\theta^2$$

Thus

$$V^2 = \frac{1}{4}\,\theta^2\bar{\theta}^2\varepsilon^{\alpha\beta}\varepsilon^{\dot{\alpha}\dot{\beta}}\sigma^m{}_{\alpha\dot{\alpha}}\sigma^n{}_{\beta\dot{\beta}}V_m V_n$$

$$= \frac{1}{4}\,\theta^2\bar{\theta}^2\sigma^m{}_{\alpha\dot{\alpha}}\sigma^{n\alpha\dot{\alpha}}V_m V_n$$

Now, we use the property

$$\sigma^m{}_{\alpha\dot{\alpha}}\sigma^{n\alpha\dot{\alpha}} = - 2\eta^{mn}$$

then

$$V^2 = - \tfrac{1}{2}\,\theta^2\bar{\theta}^2 V_m V^m \quad .$$

$V^3 = 0$ is obvious by $\theta^3 = 0$ etc.

Lecture 2: Superfield Lagrangian

A) Chiral Superfield Lagrangian:

Consider a chiral superfield, ϕ, defined by $\bar{D}_{\dot{\alpha}}\phi = 0$. The superfield ϕ can be written as

$$\phi(x,\theta,\bar{\theta}) = A(x) + \sqrt{2}\theta\psi(x) + \theta^2 F(x) + i\theta\sigma^m\,\bar{\theta}\partial_m A(x)$$

$$- \frac{i}{\sqrt{2}}\,\theta^2\partial_m\psi(x)\sigma^m\,\bar{\theta} + \frac{1}{4}\,\theta^2\bar{\theta}^2\partial^2 A(x) \tag{2.1}$$

(i) Kinetic Energy:

In the expansion of $\phi^+\phi$ we get the following component for the $\theta^2\bar{\theta}^2$ term:

$$- (\partial^n A^+)\,(\partial_n A) + i(\partial_n\bar{\psi})\sigma^n\psi + F^+F \tag{2.2}$$

It is easy to check that under the supersymmetry transformation:

$$\delta A = \sqrt{2}\,\xi\psi$$

$$\delta\psi = \sqrt{2}\,\xi\,F + i\sqrt{2}\sigma^n\,\bar{\xi}\partial_n A \tag{2.3}$$

$$\delta F = -i\sqrt{2}\,\partial_n\psi\sigma^n\bar{\xi}$$

(2.2) goes into itself plus a total divergence. Therefore, $\int d^4x\, \phi^+\phi\Big|_{\theta^2\bar\theta^2}$ is invariant under supertransformations. It follows that the kinetic energy part of the chiral superfield Lagrangian is

$$L^{\phi}_{KE} = \phi^+\phi\Big|_{\theta^2\bar\theta^2} = -(\partial^n A^+)(\partial_n A) + i(\partial_n\bar\psi)\sigma^n\psi + F^+F \tag{2.4}$$

No other component of $\phi^+\phi$ is invariant.

<u>(ii) Super Potential:</u>

(a) Consider the θ^2 component of

$$\phi = \ldots + \theta^2 F + \ldots \tag{2.5}$$

Since $\delta F = \partial_n(-i\sqrt{2}\psi\sigma^n\bar\xi)$, $\int d^4x\, \phi\Big|_{\theta^2}$ is invariant under supertransformations. No other component of ϕ is invariant.

$$\phi\Big|_{\theta^2} = F \tag{2.6}$$

This component has dimension 2.

(b) Consider the θ^2 component of ϕ^2

$$\phi^2 = \ldots + \theta^2(2AF-\psi^2) + \ldots \tag{2.7}$$

Since $\delta(2AF-\psi^2) = \partial_n(i2\sqrt{2}\,\bar\xi\sigma^n A\psi)$, $\int d^4x\, \phi^2\Big|_{\theta^2}$ is invariant under supertransformation. No other component of ϕ^2 is invariant.

$$\phi^2\Big|_{\theta^2} = 2AF - \psi^2 \tag{2.8}$$

This component has dimension 3.

(c) Consider the θ^2 component of ϕ^3.

$$\phi^3 = \ldots + \theta^2(3A^2F-3A\psi^2) + \ldots \tag{2.9}$$

Since $\delta(3A^2F-3A\psi^2) = \partial_n F^n$, $\int d^4x\, \phi^3\Big|_{\theta^2}$ is supersymmetry invariant. This term has dimension 4 and, therefore, for our theory to be renormalizable we can go no further.

$$\phi^3\Big|_{\theta^2} = 3A^2F - 3A\psi^2 \tag{2.10}$$

The superpotential $W(\phi)$ is defined as

$$W(\phi) = [f\phi + \frac{m}{2}\phi^2 + \frac{\lambda}{3}\phi^3]\Big|_{\theta^2} \tag{2.11}$$

Improvement:

We know that $\int d\theta = 0$ and $\int d\theta\ \theta = 1.$ (2.12)

Let
$$\begin{aligned} d^2\theta &= \frac{1}{4}\, d\theta^{\alpha} d\theta_{\alpha} \\ &= -\frac{1}{4}\,\varepsilon_{\alpha\beta} d\theta^{\alpha} d\theta^{\beta} \\ &= \frac{1}{4}[d\theta^1 d\theta^2 - d\theta^2 d\theta^1] \\ &= \frac{1}{2}\, d\theta^1 d\theta^2 \end{aligned} \tag{2.13}$$

Therefore $\int d^2\theta\ \theta^2 = \int \frac{1}{2}\, d\theta^1 d\theta^2 (-2\theta^1\theta^2)$

$$= \int d\theta^1\theta^1 \int d\theta^2\theta^2 = 1$$

$$\int d^2\theta\ \theta^2 = 1 \tag{2.14}$$

Similarly,

$$d^2\bar{\theta} = -\frac{1}{4}\, d\bar{\theta}_{\dot{\alpha}} d\bar{\theta}^{\dot{\alpha}}$$

$$\int d^2\bar{\theta}\ \bar{\theta}^2 = 1 \tag{2.15}$$

Also,

$$d^4\theta = d^2\theta\ d^2\bar{\theta}$$

$$\int d^4\theta\ \theta^2\bar{\theta}^2 = 1 \tag{2.16}$$

Now, if $S(x,\theta,\bar{\theta})$ is our arbitrary superfield we get

$$\int d^4\theta\ S(x,\theta,\bar{\theta}) = S(x,\theta,\bar{\theta})\Big|_{\theta^2\bar{\theta}^2} \quad \text{only.}$$

Since, for example,

$$\int d^4\theta\ \theta^2\bar{\theta}_{\dot{\alpha}}\bar{\psi}^{\dot{\alpha}} = -\int d^2\theta\ \theta^2 \int d^2\bar{\theta}\ \bar{\theta}_{\dot{\alpha}}\ \bar{\psi}^{\dot{\alpha}} = 0 \ . \tag{2.17}$$

Therefore $L_K^{\phi}{}_E$ can be written as

$$L_{K\ E}^{\phi} = \phi^+\phi\Big|_{\theta^2\bar{\theta}^2} = \int d^4\theta\ \phi^+\phi \tag{2.18}$$

Again,

$$\int d^2\theta\ S(x,\theta,\bar\theta) = S\Big|_{\theta^2} + S\Big|_{\theta^2\bar\theta} + S\Big|_{\theta^2\bar\theta^2} \qquad \text{(schematically)}.$$

However, if S is chiral, then

$$\int d^4x\ S\Big|_{\theta^2\bar\theta} = \int d^4x\ S\Big|_{\theta^2\bar\theta^2} = 0$$

because $S\Big|_{\theta^2\bar\theta}$ and $S\Big|_{\theta^2\bar\theta^2}$ are total divergences. Therefore,

$$W(\phi) = \int d^2\theta[f\theta + \frac{m}{2}\phi^2 + \frac{\lambda}{3}\phi^3] \tag{2.19}$$

For a chiral field ϕ, we get,

$$\begin{aligned} L &= \int d^4\theta\ \phi^+\phi - [\int d^2\theta(f\phi + \frac{m}{2}\phi^2 + \frac{\lambda}{3}\phi^3) + h.c.] \\ &= -(\partial^n A^+)(\partial_n A) + i(\partial_n\bar\psi)\bar\sigma^n\psi + F^+F \\ &\quad - [fF + m(AF - \frac{1}{2}\psi^2) + \lambda(A^2F - A\psi^2) + h.c.] \end{aligned} \tag{2.20}$$

We can eliminate the auxiliary field F by using its equation of motion:

$$F^+ = f + mA + \lambda A^2 \tag{2.21}$$

Substituting in L we get

$$\begin{aligned} L = &-(\partial^n A^+)(\partial_n A) + i(\partial_n\bar\psi)\sigma^n\psi \\ &-[-\frac{m}{2}\psi^2 - \lambda A\psi^2 + h.c.] \\ &- F^+F \end{aligned} \tag{2.22}$$

The potential energy $V_F = F^+F \geq 0$. Supersymmetry is not spontaneously broken if and only if the value of V_F at the minima is zero.

B) Vector Superfields and Gauge Invariant Lagrangians:

Vector superfields, V, satisfies the constraint $V = V^+$. The component field expansion is:

$$\begin{aligned} V(x,\theta,\bar\theta) = &\ C + i\theta\chi - i\bar\theta\bar\chi + \frac{i}{2}\theta^2(M+iN) - \frac{i}{2}\bar\theta^2(M-iN) \\ &- \theta\sigma^m\bar\theta\ V_m + i\theta^2\bar\theta(\bar\lambda + \frac{i}{2}\bar\sigma^m\partial_m\chi) - i\bar\theta^2\theta(\lambda - \frac{i}{2}\sigma^m\partial_m\bar\chi) \\ &+ \frac{1}{2}\theta^2\bar\theta^2(D + \frac{1}{2}\partial^2C) \quad . \end{aligned} \tag{2.23}$$

We shall consider an internal symmetry group G = U(1). If ϕ is chiral, then the action of this group on ϕ is defined as:

$$\phi \rightarrow \phi' = e^{-i\Lambda}\phi \quad .$$

Theorem: If $\phi \rightarrow \phi' = e^{-i\Lambda}\phi$, then ϕ' is chiral if Λ is chiral.

Proof:

$$\begin{aligned} \bar{D}_{\dot\alpha}\phi' &= \bar{D}_{\dot\alpha}(\phi - i\Lambda\phi + \ldots) \\ &= \bar{D}_{\dot\alpha}\phi - i[(\bar{D}_{\dot\alpha}\Lambda)\phi + \Lambda(\bar{D}_{\dot\alpha}\phi)] + \ldots \\ &= 0 \text{ iff } \bar{D}_{\dot\alpha}\Lambda = 0 \end{aligned} \tag{2.24}$$

Global action of U(1) with Λ a real, <u>constant</u> chiral superfield leaves $\phi^+\phi$ invariant.

$$\phi^{+'}\phi' = \phi^+ e^{i\Lambda^+} e^{-i\Lambda}\phi = \phi^+\phi\, e^{i(\Lambda-\Lambda)} = \phi^+\phi \tag{2.25}$$

Local action of U(1) with Λ a <u>non-constant</u> chiral superfield yields

$$\phi^{+'}\phi' = \phi^+\phi\, e^{i(\Lambda^+-\Lambda)}$$

which is not invariant. To make $\phi^+\phi$ invariant we replace $\phi^+\phi$ by $\phi^+e^{gV}\phi$ where g is a real constant. The hermiticity of $\phi^+e^{gV}\phi$ requires that $V = V^+$. Thus, V is a vector superfield. Let

$$V \rightarrow V' = V - \frac{i}{g}(\Lambda^+ - \Lambda) \text{ under the local action of U(1).}$$

Then, $\phi^{+'}e^{gV'}\phi' = \phi^+e^{gV}\phi$ is an invariant. Therefore the kinetic energy term of a chiral superfield with local U(1) invariance becomes

$$L_{K\,E} = \int d^4\theta\ \phi^+e^{gV}\phi \tag{2.26}$$

We know that in the Wess-Zumino gauge the components C, M, N and χ of superfield V have been transformed away. In this gauge we have

$$V = -\theta\sigma^m\bar{\theta}v_m + i\theta^2\bar{\theta}\bar{\lambda} - i\bar{\theta}^2\theta\lambda + \frac{1}{2}\theta^2\bar{\theta}^2 D \tag{2.27}$$

If a supersymmetry transformation is performed the W-Z gauge can be restored by a gauge transformation. After fixing W-Z gauge there is still a subset of gauge transformations such that

$$\begin{aligned} V_m &\rightarrow V_m - g^{-1}\partial_m(A_\Lambda + A_\Lambda^+) \\ \lambda &\rightarrow \lambda \\ D &\rightarrow D \end{aligned} \tag{2.28}$$

This implies that to quantize the theory in W-Z gauge we must still fix the gauge ambiguity of V_m with the usual Fadeev-Popov procedure. Since in this gauge $V^n = 0$ for $n \geq 3$, the expansion of e^{gV} terminates and we find that

$$\int d^4\theta\ \phi^+ e^{gV}\phi = -(D^n A^+)(D_n A) + i(D_n\bar{\psi})\sigma^n\psi + F^+F - \frac{ig}{\sqrt{2}}(\bar{\lambda}\bar{\psi}\ A - \lambda\psi A^+) + \frac{1}{2}DA^+A \qquad (2.29)$$

where $D_n = \partial_n + \frac{ig}{2} V_n$.

We emphasize that (2.29) is valid only in Wess-Zumino gauge.

Superpotential $W(\phi)$ is locally, as well as globally, U(1) invariant and does not need to be modified. However, we still need the kinetic energy for vector superfield V. We must find a superfield which is invariant under $V \rightarrow V' = V - \frac{i}{g}(\Lambda^+ - \Lambda)$. Let us consider

$$W_\alpha = -\frac{1}{4}\bar{D}^2 D_\alpha V \qquad (2.30)$$

W_α is gauge invariant

$$\begin{aligned} W'_\alpha &= -\frac{1}{4}\bar{D}^2 D_\alpha[V - \frac{i}{g}(\Lambda^+ - \Lambda)] \\ &= W_\alpha - \frac{i}{4g}\bar{D}^2 D_\alpha \Lambda \\ &= W_\alpha + \frac{i}{4g}\bar{D}^{\dot{\alpha}}\{\bar{D}_{\dot{\alpha}}, D_\alpha\}\Lambda \\ &= W_\alpha + \frac{i}{4g}\bar{D}^{\dot{\alpha}} 2\sigma^m_{\alpha\dot{\alpha}} P_m\Lambda \\ &= W_\alpha \ \{\text{Since, } [\bar{D}^{\dot{\alpha}}, P_m] = 0\} \end{aligned} \qquad (2.31)$$

W_α is also chiral, since

$$\bar{D}_{\dot{\alpha}} W_\alpha \propto \bar{D}_{\dot{\alpha}}\bar{D}^2 D_\alpha V = 0 \qquad (2.32)$$

This implies

$$L^V_{K\,E} = \frac{1}{4}\int d^2\theta\ W^\alpha W_\alpha + \text{h.c.} \qquad (2.33)$$

$L^V_{K\,E}$ is gauge invariant and is most easily evaluated in the W-Z gauge. In component fields $L^V_{K\,E}$ is given by

$$L^V_{K\,E} = -\frac{1}{4}V_{mn}V^{mn} - i\lambda\sigma^m\partial_m\bar{\lambda} + \frac{1}{2}D^2 \qquad (2.34)$$

where $V_{mn} = \partial_m V_n - \partial_n V_m$. There is also another supersymmetry invariant.

Consider

$$\int d^4\theta V' = \int d^4\theta[V - \frac{i}{g}(\Lambda^+ - \Lambda)]$$

$$= \int d^4\theta V - \frac{i}{g}\int d^4\theta(\Lambda^+ - \Lambda) \qquad (2.35)$$

Since Λ is chiral, $\int d^4x\, d^4\theta(\Lambda^+ - \Lambda)$ vanishes. Therefore, we get

$$L_D = 2\kappa\int d^4\theta V \qquad (2.36)$$

is an invariant, which, in the W-Z gauge, is simply

$$L_D = \kappa D \qquad (2.37)$$

Now, putting everything together we get, for arbitrary gauge,

$$L = [\frac{1}{4}\int d^2\theta W^\alpha W_\alpha + \text{h.c.}] + 2\kappa\int d^4\theta V$$

$$+ \int d^4\theta\phi^+ e^{gV}\phi - [\int d^2\theta W(\phi) + \text{h.c.}] \qquad (2.38)$$

In component form, we get in W-Z gauge

$$L = -\frac{1}{4} V_{mn}V^{mn} - i\lambda\sigma^m\partial_m\bar{\lambda} + \frac{1}{2}D^2 + \kappa D$$

$$-(D^n A^+)(D_n A) + i(D_n\bar{\psi})\sigma^n\psi + F^+F$$

$$-\frac{ig}{\sqrt{2}}(\bar{\lambda}\bar{\psi}A - \lambda\psi A^+) + \frac{1}{2} g\, D\, A^+A + \ldots. \qquad (2.39)$$

We can use the equation of motion to eliminate the auxiliary field D.

$$D = -\kappa - \frac{g}{2} A^+A \qquad (2.40)$$

Substituting into L we get

$$L = L(\text{no } D) - \frac{1}{2} D^2 \quad . \qquad (2.41)$$

Thus, we get the vector field contribution to the potential energy $V_D = \frac{1}{2} D^2 \geq 0$. Eliminating F as before,

$$L = L(\text{no D or F}) - V_T$$

$$V_T = F^+F + \frac{1}{2} D^2 \qquad (2.42)$$

This expression determines the total potential energy of the system which determines the vacuum.

C) Spontaneous Supersymmetry Breaking:

From the anticommutator

$$\{\hat{Q}_\alpha, \hat{\bar{Q}}_{\dot{\alpha}}\} = 2\,\sigma^m_{\alpha\dot{\alpha}}\,\hat{P}_m \tag{2.43}$$

it follows that

$$\hat{P}_o = \hat{H} = \frac{1}{4}\sum_{i=1}^{2}(\hat{Q}_i\hat{\bar{Q}}_i + \hat{\bar{Q}}_i\hat{Q}_i) \tag{2.44}$$

Supersymmetry is spontaneously broken if and only if

$$\hat{Q}_\alpha\,|0> \neq 0 \tag{2.45}$$

It is clear from Eqn. (2.44) that this condition holds if and only if

$$<0|\hat{H}|0> > 0 \Rightarrow <V_T> > 0 \tag{2.46}$$

We conclude that supersymmetry is spontaneously broken if and only if

$$<F> > 0 \text{ and/or } <D> > 0 \tag{2.47}$$

The supersymmetry transformations of chiral and gauge fermions are given by

$$\delta\psi = i\sqrt{2}\,\sigma^n\,\bar{\xi}\partial_n A + \sqrt{2}\,\xi\,F$$

$$\delta\lambda = \sigma^{mn}\,\xi\,V_{mn} + i\xi D \tag{2.48}$$

respectively. Expanding the fields around the translationally invariant state implies that

$$\delta\psi = \sqrt{2}\,\xi\,<F>$$

$$\delta\lambda = i\xi\,<D> \tag{2.49}$$

It follows that supersymmetry is spontaneously broken if and only if $\delta\psi$ and/or $\delta\lambda$ transform inhomogeneously. Note that such inhomogeneous transformations imply the existence of a zero mass fermion in the spectrum - the Goldstone fermion.

Lecture 3: Local Supersymmetry (Supergravity)

A) Space-Time:

a) Global Lorentz Symmetry $\Longrightarrow$ Flat Space

b) Local Lorentz Symmetry $\Longrightarrow$ Curved Space (Gravity)

B) Superspace:

a) Global Supersymmetry $\Longrightarrow$ Flat Superspace (with torsion)

b) Local Supersymmetry $\Longrightarrow$ Curved Superspace (Supergravity)

(i) Scalar Curvature Chiral Superfield:

The scalar curvature superfield is given by

$$
\begin{aligned}
R = & -\frac{1}{6}M + \theta(\sigma^a\sigma^b\psi_{ab} - i\sigma^a\bar{\psi}_a M + i\psi_a b^a) \\
& + \theta^2\Big(-\frac{1}{2}\mathcal{R} + i\bar{\psi}^a\sigma^b\psi_{ab} + \frac{2}{3}M^\dagger M + \frac{1}{3}b^a b_a \\
& - i\, e_a{}^m \mathcal{D}_m b^a + \frac{1}{2}\bar{\psi}\bar{\psi}M - \frac{1}{2}\psi_a\sigma^a\bar{\psi}_b b^b \\
& + \frac{1}{8}\varepsilon^{abcd}[\bar{\psi}_a\sigma_b\psi_{cd} + \psi_a\sigma_b\bar{\psi}_{cd}]\Big)
\end{aligned}
\tag{3.1}
$$

where indices from the beginning (middle) of the alphabet, such as a, b (m,n), refer to the tangent space (manifold). The component fields in R are

$$
\begin{aligned}
& e_a{}^m \text{ - graviton, spin 2} \\
& \psi_a{}^\alpha \text{ - gravitino, spin 3/2} \\
& ba \text{ - auxiliary field, spin 1} \\
& M \text{ - auxiliary field, spin 0}
\end{aligned}
\tag{3.2}
$$

$\mathcal{R}$ and $\mathcal{D}_m$ in Eqn. (3.1) are functions of $e_a{}^m$ and $\psi_a{}^\alpha$. The local supersymmetry transformations for the component fields in (3.2) can be calculated. They show that $e_a{}^m$, $\psi_a{}^\alpha$, b_a, and M transform into one another irreducibly.

(ii) Chiral Density Superfield:

The chiral density superfield is given by

$$
\mathcal{E} = e + \theta(ie\sigma^a\bar{\psi}_a) - \theta^2 e(M^\dagger + \bar{\psi}_a\sigma^{ab}\bar{\psi}_b)
\tag{3.3}
$$

where $e = \det e_a{}^m$. Note that $\mathcal{E}$ has the same component fields (excluding b_a) as R. Clearly, $\mathcal{E}$ is the generalization of the usual density $\sqrt{-g}$ to superspace. That is

$$
\int d^2\theta \rightarrow \int d^2\,\theta\,\mathcal{E}
\tag{3.4}
$$

in curved superspace.

(iii) Supergravity Lagrangian:

The supergravity Lagrangian is given by

$$L_{SG} = -6\, M_p^2 (\int d^2\theta\, \mathcal{E}\, R + h.c.) \tag{3.5}$$

where M_p is the Planck mass. In component fields

$$L_{SG} = M_p^{\,2} \{ -\frac{1}{2}\, e\mathcal{R} - \frac{1}{3} e M^\dagger M + \frac{1}{3}\, e b_a b^a$$

$$+ \frac{1}{2}\, e\, \varepsilon^{klmn} (\bar{\psi}_k \sigma_1 \tilde{\mathcal{D}}_m \psi_n - \psi_k \sigma_1 \tilde{\mathcal{D}}_m \bar{\psi}_n) \} \tag{3.6}$$

where

$$\mathcal{R} = e_a^{\ n}\, e_b^{\ m} (\partial_n \omega_m^{\ ab} - \partial_m \omega_n^{\ ab} + \omega_m^{\ ac} \omega_{nc}^{\ \ b} - \omega_n^{\ ac} \omega_{mc}^{\ \ b})$$

$$\omega = \text{function of } e_a^{\ m},\ \psi_a^{\ \alpha} \tag{3.7}$$

$$\tilde{\mathcal{D}}_m \psi_n^{\ \alpha} = \partial_m \psi_n^{\ \alpha} + \psi_n^{\ \beta} \omega_{m\beta}^{\ \ \ \alpha}$$

Now rescale the component fields as follows:

$$\begin{aligned} e_a^{\ m} &\to e_a^{\ m'} = M_p\, e_a^{\ m}, && \text{dimension } 1 \\ \psi_a^{\ \alpha} &\to \psi_a^{\ \alpha'} = M_p \psi_a^{\ \alpha}, && \text{dimension } 3/2 \\ b_a &\to b_a^{\ '} = M_p b_a, && \text{dimension } 2 \\ M &\to M' = M_p M, && \text{dimension } 2 \end{aligned} \tag{3.8}$$

Dropping the prime notation, the supergravity Lagrangian becomes

$$L_{SG} = -\frac{1}{2}\, e\mathcal{R} - \frac{1}{3} e M^\dagger M + \frac{1}{3} e\, b^a b a$$

$$+ \frac{1}{2}\, e\varepsilon^{klmn} (\bar{\psi}_k \sigma_1 \tilde{\mathcal{D}}_m \psi_n - \psi_k \sigma_1 \tilde{\mathcal{D}}_m \bar{\psi}_n) \tag{3.9}$$

The equations of motion of auxiliary fields M and b_a are

$$\begin{aligned} M &= 0 \\ b^a &= 0 \end{aligned} \tag{3.10}$$

It follows that the supergravity Lagrangian is

$$L_{SG} = -\frac{1}{2}\, e\,\mathcal{R} + \frac{1}{2}\, e\varepsilon^{klmn} (\bar{\psi}_k \sigma_1 \tilde{\mathcal{D}}_m \psi_n - \psi_k \sigma_1 \tilde{\mathcal{D}}_m \bar{\psi}_n) \tag{3.11}$$

The first and second terms are the Einstein and Rarita-Schwinger

Lagrangians respectively.

(iv) Supergravity Coupled to Matter:

In flat superspace the most general matter Lagrangian is

$$L = \frac{1}{4}(\int d^2\theta W^\alpha W_\alpha + h.c.) + \int d^4\theta\ \phi_i^{\dagger} e^{gV} \phi_i$$

$$+ (\int d^2\theta\ [a_i\phi_i + \frac{1}{2} m_{ij}\phi_i\phi_j + \frac{1}{3}\lambda_{ijk}\phi_i\phi_j\phi_k] + h.c.) \qquad (3.12)$$

Note that

$$\int d^4\theta = \int d^2\theta(-\frac{1}{4}\bar{D}^2) \qquad (3.13)$$

Therefore the Lagrangian can be written as

$$L = (\int d^2\theta[\frac{1}{4}W^\alpha W_\alpha - \frac{1}{8}\bar{D}^2\ \phi_i^{\dagger}\ e^{gV}\phi_i + a_i\phi_i$$

$$+ \frac{1}{2} m_{ij}\phi_i\phi_j + \frac{1}{3}\lambda_{ijk}\phi_i\phi_j\phi_k] + h.c.) \qquad (3.14)$$

The conversion of this Lagrangian to curved superspace can now be accomplished using the following rules

$$\int d^2\theta \to \int d^2\theta\ \mathcal{E}$$

$$\bar{D}^2 \to \bar{\mathcal{D}}^2 - 8R \qquad (3.15)$$

$\bar{\mathcal{D}}$ is the covariant derivative in curved superspace. Therefore, in curved superspace, the above Lagrangian becomes

$$L = (\int d^2\theta\ \ [\frac{1}{4}W^\alpha W_\alpha - \frac{1}{8}(\bar{\mathcal{D}}^2 - 8R)\ \frac{\phi_i^{\dagger} e^{gV}\phi_i}{M_p^2} +$$

$$c + a_i\phi_i + \frac{1}{2} m_{ij}\phi_i\phi_j + \frac{1}{3}\lambda_{ijk}\phi_i\phi_j\phi_k] + h.c.) \qquad (3.16)$$

$$+ (\int d^2\theta\ \mathcal{E}(-6R) + h.c.)$$

Note that we can write

$$\int d^2\theta\ \mathcal{E}(-6R) + \int d^2\theta\ [-\frac{1}{8}(\bar{\mathcal{D}}^2 - 8R)\ \frac{\phi_i^{\dagger} e^{gV}\phi_i}{M_p^2}] =$$

$$= \frac{3}{4}\int d^2\theta\ \mathcal{E}(\bar{\mathcal{D}}^2 - 8R)\ [1 - \frac{1}{6}\frac{\phi_i^{\dagger} e^{gV}\phi_i}{M_p^2}] \qquad (3.17)$$

It is clear that this term can be generalized to

$$\frac{3}{4}\int d^2\theta\, \mathcal{E}(\bar{\mathcal{D}}^2 - 8R)\, e^{-\frac{1}{6}\frac{\phi_i^\dagger e^{gV}\phi_i}{M_p^2}} \tag{3.18}$$

In fact, a further generalization to

$$\frac{3}{4}\int d^2\theta\, \mathcal{E}\,(\bar{\mathcal{D}}^2 - 8R)\, e^{-\frac{1}{6}K} \tag{3.19}$$

where K is an arbitrary function of $\phi_i^\dagger e^{gV}\phi_i/M_p^2$ is permitted. Note that taking

$$K(x) = -6\,\ell n\,(1 - \frac{1}{6}x),\quad x = \frac{\phi_i^\dagger e^{gV}\phi_i}{M_p^2} \tag{3.20}$$

reproduces the original Lagrangian. It follows that the most general Lagrangian in curved superspace is

$$L = (\int d^2\theta\mathcal{E}[\frac{1}{4}W^\alpha W_\alpha + \frac{3}{4}(\bar{\mathcal{D}}^2 - 8R)\, e^{-\frac{1}{6}K\frac{(\phi_i^\dagger e^{gV}\phi_i)}{M_p^2}}$$

$$+ W(\phi_i)] + \text{h.c.}) \tag{3.21}$$

where

$$W(\phi_i) = c + a_i\phi_i + \frac{1}{2}m_{ij}\phi_i\phi_j + \frac{1}{3}\lambda_{ijk}\phi_i\phi_j\phi_k \tag{3.22}$$

In component fields this Lagrangian is very complicated. However, the important new physics implied by this Lagrangian is easy to deduce. Consider the couplings of $e_a{}^m$, $\psi_a{}^\alpha$, b_a, and M to the scalar component A_i of superfield ϕ_i. The Feynman graphs of the $e_a{}^m$, $\psi_a{}^\alpha$, and b_a couplings are

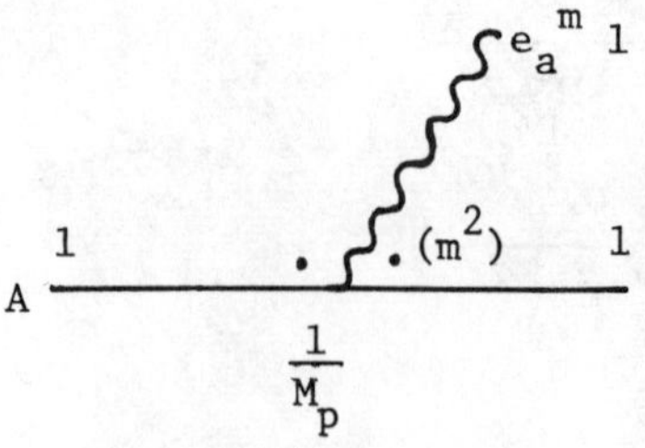

,
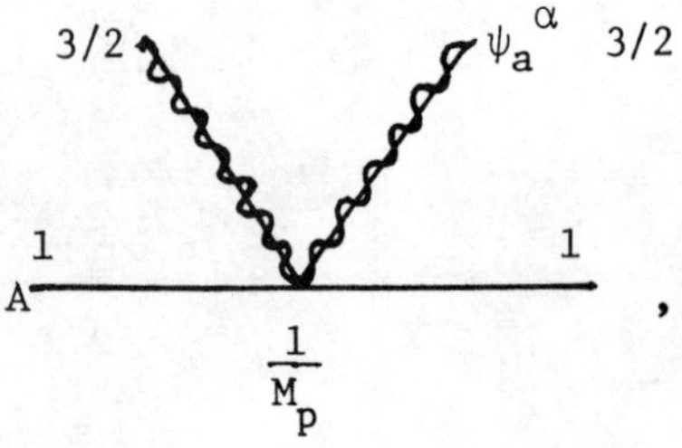

,

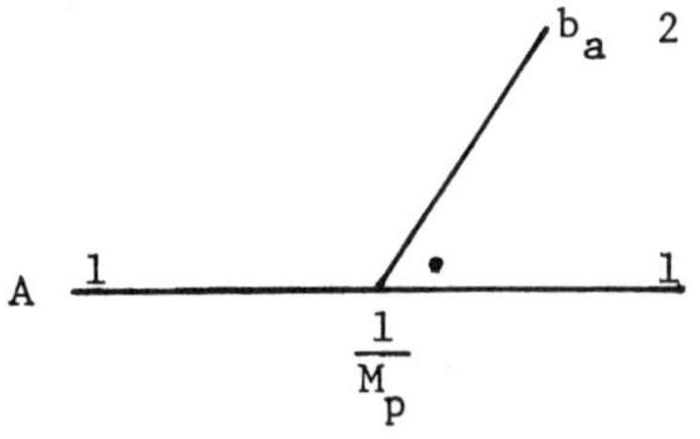

Therefore, these interactions are suppressed by m/M_p or p/M_p. The coupling of M to A_i can be represented as

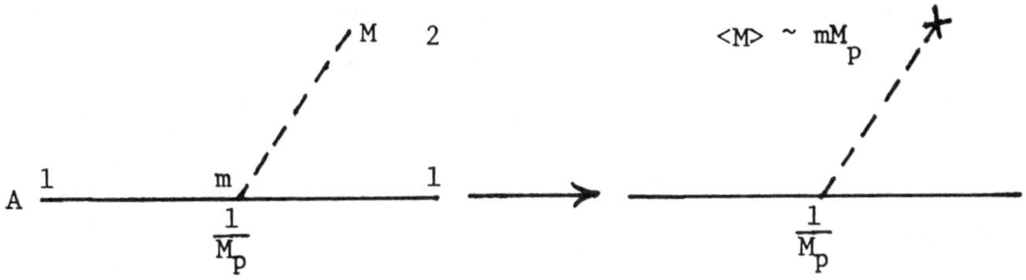

If $\langle M\rangle \sim mM_p$, then the M interaction is not naturally suppressed.

As we will see, $\langle M\rangle$ is of this order of magnitude. We conclude that supergravity effects low energy physics through the M field. Eliminate all auxiliary fields (F, D, b_a, M) using their equations of motion. Note that

$$M = -\frac{3}{M_p} W(A_i) + \ldots \tag{3.23}$$

Define the Kahler potential, Kahler metric, and Kahler covariant derivative to be

$$K = K\,(A_i^{\dagger} A_i)$$

$$g_i^{\ j*} = \frac{\partial^2 K}{\partial A_i \partial \bar{A}_{j*}}$$

$$D_{A_i} \mathcal{W}(A) = \frac{\partial \mathcal{W}}{\partial A_i} + \frac{\partial K}{\partial A_i} \mathcal{W} \tag{3.24}$$

respectively, where $\mathcal{W}$ is any function of A_i. Then after eliminating the auxiliary fields, Lagrangian (3.21) becomes (in component fields)

$$L = -\frac{1}{2} e \mathcal{R} - \frac{1}{2} e \varepsilon^{klmn} (\bar{\psi}_k \sigma_l \tilde{\mathcal{D}}_m \psi_n - \psi_k \sigma_l \tilde{\mathcal{D}}_m \bar{\psi}_n)$$

$$- e g_i^{\ j*} (\mathcal{D}_m A_i)(\mathcal{D}^m A_j)^\dagger + i e g_i^{\ j*} (\mathcal{D}_m \bar{\psi}_{j*}) \sigma^m \psi_i$$

$$- \frac{1}{4} e V_{mn} V^{mn} - i e \lambda \sigma^m \partial_m \bar{\lambda} \qquad (3.25)$$

+ interactions with fermions

$$- V(A_i)$$

where

$$V(A_i) = e^K [g_i^{\ j*} (D_{A_i} W)(D_{A_j} W)^\dagger - \frac{3}{M_p^{\ 2}} |W|^2] \qquad (3.26)$$

$$+ \frac{1}{2} D^2$$

is the potential energy. Note that V is no longer non-negative. Therefore, one can adjust the parameters in W such that

$$\langle V \rangle = 0 \Longrightarrow \text{zero cosmological constant} \qquad (3.27)$$

Lagrangian (3.25) is invariant under local supertransformations that include

$$\delta\psi_i = (\text{fermions and } \partial_m A_j) + \sqrt{2}\, e^{K/2} g_i^{\ j*} (D_{A_j} W)^\dagger \xi$$

$$\delta\lambda = (\text{fermions and } \partial_m V_n) + i D \xi$$

It follows that there is a Goldstone fermion if and only if

$$\langle \sqrt{2}\, e^{K/2} g_i^{\ j*} (D_{A_j} W)^\dagger \rangle \neq 0 \text{ and/or } \langle D \rangle \neq 0 \qquad (3.29)$$

Therefore, supersymmetry is spontaneously broken if and only if

$$\langle D_{A_j} W \rangle \neq 0 \text{ and/or } \langle D \rangle \neq 0 \qquad (3.30)$$

Note that $\langle D_{A_j} W \rangle \neq 0$ and $\langle V \rangle = 0$ together imply that

$$\langle W \rangle \neq 0 \Longrightarrow \langle M \rangle \neq 0 \qquad (3.31)$$

The gravitino mass can then be shown to be

$$m_{3/2} \sim |\langle W \rangle| / M_p^{\ 2} \qquad (3.32)$$

It follows from Eqns. (3.23) and (3.32) that

$$\langle M \rangle \sim m_{3/2} M_p \qquad (3.33)$$

We conclude that if $m_{3/2} \gtrsim m_{EW}$ then the effect of the M field on low energy physics is substantial.

Lecture 4: Supergravity Grand Unified Theories

Simplest Possible Theory:

A) Gauge Group:

$$G = SU(5)$$

B) Chiral Superfields:

1) Σ. Σ is a 24 under SU(5).

The scalar component of Σ gets a vacuum expectation value (VEV) which breaks

$$SU(5) \to SU(3)_C \times SU(2)_L \times U(1)_Y$$

The associated Kahler derivative, $D_\Sigma W$, gets a VEV which breaks supersymmetry. Note that SU(5) and supersymmetry are broken in the same chiral superfield.

2) H, H'. H and H' are a 5 and $\bar{5}$ under SU(5).

The $SU(3)_C$ triplet component fermions mediate nucleon decay. The $SU(2)_L$ doublet component scales get VEV's which break $SU(2)_L \times U(1)_Y$ to $U(1)_{EM}$.

3) M_J, M'_J. M_J, M'_J are $\bar{5}$ and 10 under SU(5).

These are the usual lepton and quark families (J = 1,3).

4) θ, θ'. θ and θ' are 50 and $\overline{50}$ under SU(5)

The superfields produce natural $SU(3)_C$ triplet/$SU(2)_L$ doublet mass splitting in H and H'.

5) 10, 10'. 10 and 10' are 10 and $\overline{10}$ under SU(5).

The $SU(3)_C \times SU(2)_L$ singlet components of these superfields reduce M_G, $\alpha(M_G)$, and $\sin^2\theta_W$ to acceptable values.

6) ψ, ψ'. ψ, ψ' are 40 and $\overline{40}$ under SU(5).

These superfields produce natural $SU(3)_C$ triplet/singlet mass splitting in 10, 10'.

C) Superpotential:

The superpotential is given by

$$W = W_\Sigma + W_\theta + W_Y + W_\psi \tag{4.1}$$

where

$$W_\Sigma = \beta m\, M^2 + \frac{m^2}{2}\,\mathrm{Tr}\,\Sigma^2 + \frac{\lambda_1}{3}\,\mathrm{Tr}\,\Sigma^3 + \frac{\lambda_4}{4M}\,\mathrm{Tr}\,\Sigma^4 \tag{4.2}$$

$$W_\theta = M\theta'\theta + \frac{\lambda_2}{M}\,\theta'\Sigma^2 H + \frac{\lambda_3}{M}\,\theta\Sigma^2 H' \tag{4.3}$$

$$W_Y = f_{IJ} H M_I M_J + g_{IJ}\, H' M_I M'_J \tag{4.4}$$

$$W_\psi = M\psi'\psi + \frac{\lambda_5}{M}\,\psi'\Sigma^2 10 + \frac{\lambda_6}{M}\,\psi\Sigma^2 10' \tag{4.5}$$

where β, λ_i, f_{IJ}, g_{IJ} are dimensionless and m, M have dimension one.

D) The Vacuum State:

First consider W alone. Using Eqn. (3.26) calculate the potential energy, V. The potential looks like

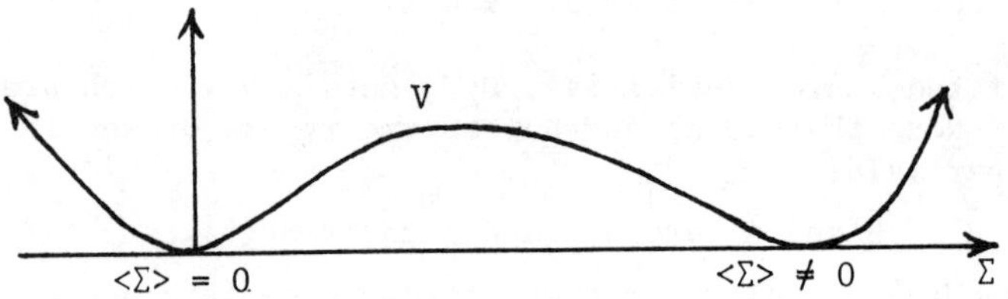

In this lecture we will ignore the $\langle\Sigma\rangle = 0$ vacuum. The non-vanishing vacuum state is found to be

$$\langle\Sigma\rangle = \frac{1}{\sqrt{10}}\, M \begin{pmatrix} 2 & & & & \\ & 2 & & & \\ & & 2 & & \\ & & & -3 & \\ & & & & -3 \end{pmatrix} \tag{4.6}$$

Clearly this VEV breaks SU(5) spontaneously down to $SU(3)_C \times SU(2)_L \times U(1)_Y$. The Kahler derivative evaluated at $\langle\Sigma\rangle$ is found to be

$$\langle D_\Sigma W\rangle = \frac{1}{3\sqrt{10}}\, mM \begin{pmatrix} 2 & & & & \\ & 2 & & & \\ & & 2 & & \\ & & & -3 & \\ & & & & -3 \end{pmatrix} \tag{4.7}$$

It follows from the discussion in Lecture 3 that this VEV breaks supersymmetry with a scale of order $\sqrt{m\,M}$. Note that this theory is not an O'Raifeartaigh model. Therefore, in the global sueprsymmetry limit $(M \to \infty)$ there would be no supersymmetry breaking. It follows that the breaking of supersymmetry, and the scale of this breaking, is a purely gravitational effect. That is, supergravity allows a new mechanism for supersymmetry breaking.

E) Naturally Heavy Higgs Color Triplets:

Now include the contribution of W_θ to the superpotential. Expand Σ around $\langle\Sigma\rangle$. The quadratic part of V involving θ, θ', H, and H' only is (to leading order)

$$M^2(\hat{\theta}^\dagger\hat{\theta} + \hat{\theta}'^\dagger\hat{\theta}') + (\theta_3^\dagger, H_3^\dagger)\mathcal{M}^2\begin{pmatrix}\theta_3\\H_3\end{pmatrix} + (\theta_3'^\dagger, H_3'^\dagger)\mathcal{M}'^2\begin{pmatrix}\theta_3'\\H_3'\end{pmatrix} \tag{4.8}$$

where

$$\mathcal{M}^2 = M^2\begin{pmatrix}8 + \frac{25}{2}\lambda_3^2 & 10\sqrt{8}\,\lambda_2\\ 10\sqrt{8}\,\lambda_2 & 100\,\lambda_2^2\end{pmatrix} \tag{4.9}$$

and $\hat{\theta}$ means everything in θ except θ_3. $\mathcal{M}'^2$ is the same as $\mathcal{M}^2$ with λ_2 and λ_3 exchanged. Note that the masses of H_2 and H_2' vanish to leading order since they can find no partner in θ and θ' to form a mass term with. Therefore, this is called the method of "missing multiplets". For simplicity, choose $\lambda_2 = \lambda_3 = 1$. Therefore

$$\mathcal{M}_D = \mathcal{M}_D' = \begin{pmatrix}3.39\,M & 0\\ 0 & 3.39M\end{pmatrix} \tag{4.10}$$

where the subscript D means diagonal. It follows that the missing multiplet method naturally splits the masses of H_2, H_2' from H_3, H_3'. H_2, H_2' are light and H_3, H_3' are heavy with masses of $O(M)$. Their fermionic superpartners have similar masses. The dominant nucleon decay mode in this theory is

$$p(n) \rightarrow K^+(K^o)\bar{\nu}_\mu \tag{4.11}$$

This decay is mediated by H_3, H_3' Higgsinos. For typical values of the parameters the nucleon lifetime is

$$\tau_n \simeq 1.6 \times 10^{-4}\, m_{H_3}^2 \tag{4.12}$$

For the masses in Eqn. (4.10)

$$\tau_n \simeq 1.06 \times 10^{34} \text{ yrs.} \tag{4.13}$$

Therefore, in this theory the nucleon lifetime is naturally long. Of course, τ_n can be reduced by choosing λ_2 and/or λ_3 to be much smaller then unity. Naturally, however, this theory predicts longer nucleon lifetimes than non-supersymmetric grand unified theories.

F) The Low Energy Effective Lagrangian:

Now include the contribution of W_Y to the potential energy. Expand around $\langle\Sigma\rangle$ and $\langle D_\Sigma W\rangle$. Then the low energy effective potential is found to be

$$V_{LE} = V_{LE}^{(1)}(y_a) + V_{LE}^{(2)}(\Sigma_3, \Sigma_8) + \frac{1}{2} D_\alpha(y,\Sigma)^2 \tag{4.14}$$

where y_a are the usual fields of the supersymmetric Weinberg-Salam model, and Σ_3 and Σ_8 are the $SU(3)_c \times SU(2)_L \times U(1)_Y$ (1,3,0) and (8,1,0) components of Σ respectively. The first term in Eqn. (4.14) is

$$V_{LE}^{(1)}(y_a) = \left|\frac{\partial \tilde{g}}{\partial y_a}\right|^2 + m_{3/2} A(\tilde{g} + h.c.) + m_{3/2}^2 |y_a|^2 \tag{4.15}$$

where

$$\tilde{g} = e^{3/2}(\lambda_{IJ}^U Q_{LI} H_2 \bar{u}_{RJ} + \lambda_{IJ}^D Q_{LI} H_2' \bar{d}_{RJ} + \lambda_{IJ}^D L_{LJ} H_2' \bar{e}_{RI})$$

$$A = 3 \tag{4.16}$$

and the gravitino mass is

$$m_{3/2} = e^{3/2} \frac{m}{3} \tag{4.17}$$

The first term in Eqn. (4.15) is supersymmetric and does not involve gravitation. The two terms proportional to $m_{3/2}$ are induced by gravity and explicitly break supersymmetry. The second term in Eqn. (4.14) is

$$V_{LE}^{(2)}(\Sigma_3,\Sigma_8) = Tr\left|\frac{\partial \tilde{f}_3}{\partial \Sigma_3}\right|^2 + Tr\left|\frac{\partial \tilde{f}_8}{\partial \Sigma_8}\right|^2 + m_{3/2} A_3(\tilde{f}_3 + h.c.)$$

$$+ m_{3/2} A_8(\tilde{f}_8 + h.c.) + m_{3/2}^2(Tr|\Sigma_3|^2 + Tr|\Sigma_8|^2) \tag{4.18}$$

where

$$\tilde{f}_3 = - m_{3/2}\left(\frac{135}{14}\right) Tr\, \Sigma_3^2$$

$$\tilde{f}_8 = m_{3/2}\left(\frac{185}{14}\right) Tr\, \Sigma_8^2 \tag{4.19}$$

$$A_3 = \frac{124}{45}, \quad A_8 = \frac{558}{185}$$

The first two terms in Eqn. (4.18) are supersymmetric, whereas the terms proportional to $m_{3/2}$ are induced by gravity and explicitly break supersymmetry. We conclude that

1) all scalars get gravity induced mass of order $m_{3/2}$ at tree level;

2) there are light $SU(3)_C$ $\underline{8}$ and $SU(2)_L$ $\underline{3}$ chiral superfields with masses of order $m_{3/2}$;

3) supersymmetry is explicitly broken in the low energy Lagrangian;

4) the effective Lagrangian is renormalizable.

Finally, calculate M_G, $\alpha(M_G)$, and $\sin^2\theta_W$ to the one-loop level. Without 10 and 10' superfields all of these parameters are found to be too large. Now include the contribution of W_ψ to the potential energy. The $SU(3)_C \times SU(2)_L \times U(1)_Y$ $(\bar{3}, 1, -2/3)$, $(3, 2, 1/6)$ components of 10 find partners in ψ' and get masses of order M. There is, however, no component of ψ' to which the (1,1,1) component of 10 can couple quadratically. The same is true for 10'. Therefore, the missing multiplet method naturally splits the masses of the $SU(3)_C$ triplets from the singlets in 10 and 10'. The singlets are light, and the triplets heavy with masses of order M. If we take $m_{3/2}$ = 250 GeV, $\alpha_3(10^2\ \text{GeV}) = 1/10$, and $\alpha_{EM}(10^2\ \text{GeV}) = 1/137$ then we find that, at the one-loop level,

$$M_G = 3.43 \times 10^{18}\ \text{GeV}$$

$$\alpha(M_G) = .082$$

$$\sin^2\theta_W\ (10^2\ \text{GeV}) = .226$$

which are acceptable values.

REFERENCES

1. For the subjects discussed in Lectures 1,2, and 3 we refer the reader to Supersymmetry and Supergravity, by J. Wess and J. Bagger, Princeton University Press (1983), and references therein.

2. For the subjects discussed in Lecture 4 we refer the reader to B. Ovrut and S. Raby, Phys. Lett. 138B (1984) 72-76, and references therein.

COMPOSITE MODELS OF "ELEMENTARY" PARTICLES AND FIELDS

H. Terazawa
Institute for Nuclear Study, University of Tokyo
Midori-cho, Tanashi, Tokyo 188, Japan

ABSTRACT

Composite models of "elementary" particles and fields, leptons, quarks, and gauge bosons (and also Higgs scalars, if any), are reviewed and discussed in a pedagogical way.

CONTENTS

I. Introductory Review of Composite Models
II. Unified and Grand Unified Composite Models
III. Supergrand Unified Composite Model
IV. Concluding Remark

I. INTRODUCTORY REVIEW OF COMPOSITE MODELS

As it stands now, high energy physics may be best expressed in a word, "lepton-quark physics". There exist at least six leptons and five flavors and three colors (i=1,2,3) of quarks, altogether, twenty-one leptons and quarks:

$$\begin{pmatrix} \nu_e \\ e \end{pmatrix} \begin{pmatrix} \nu_\mu \\ \mu \end{pmatrix} \begin{pmatrix} \nu_\tau \\ \tau \end{pmatrix}$$

$$\begin{pmatrix} u_i \\ d_i \end{pmatrix} \begin{pmatrix} c_i \\ s_i \end{pmatrix} \begin{pmatrix} \\ b_i \end{pmatrix} .$$

In some models, more leptons and more quarks are expected to be found. It is now hard to believe that all of these leptons and quarks are elementary. It seems to me that we must believe the existence of more fundamental particles, the subquarks, which are building-blocks of these leptons and quarks. In fact, for the last several years, many subquark models have been proposed and discussed.[1] In this chapter, I would like to present an introductory review of subquark models and to present some important problems for future investigation. A similar review has been presented at the 1981 INS Symposium on Quark and Lepton Physics which was held in 1981.[2] It must also be mentioned that regrettably I have no time to review the long history of related works starting from the "Nagoya model" in 1960, which I have presented in Ref. 1.

The subjects which I shall cover in this chapter include the following:

1. Origin of Flavors and Colors

0094-243X/86/1430122-49$3.00

2. Lepton and Quark Mass Spectra
3. Quark (and Lepton) Mixings
4. Lepton and Quark Magnetic Moments
5. Leptons and Quarks of Spin 3/2 or Higher
6. Exotic States of Subquarks
7. Generation (or Family) Problem
8. Techniquarks vs. Subquarks
9. Grand Unification vs. Subquarks
10. Subquark Supergravity vs. Subquark Pregeometry
11. Conclusion.

Before going into these individual subjects, let me quickly review what we proposed in 1976 as a standard of reference.

In our spinor-subquark model,[3] there exist 6+N (N=1,2,3,···) subquarks of spin 1/2,

$$\begin{pmatrix} w_1^{+1/2} \\ w_2^{-1/2} \end{pmatrix} \quad (h_1^{\,0}\ h_2^{\,0}\ h_3^{\,0} \cdots h_N^{\,0}) \quad (C_0^{-1/2}\ C_1^{+1/6}\ C_2^{+1/6}\ C_3^{+1/6}).$$

The w_i's (i=1,2) form a doublet of the weak isospin of $SU(2)_w$ and are called "wakems" as they are related with the weak and electromagnetic interactions. The h_i's ($i=1 \sim N$) form a N-plet of the horizontal gauge symmetry of $SU(N)_h$[4] and are called "hakams". The C_0 and C_i's (i=1,2,3) form a singlet and triplet of color $SU(3)_C$ and called "chroms".[5] The most suitable charge assignment for these subquarks is given by $Q = \pm 1/2$ for w_i, $Q = 0$ for h_i, $Q = -1/2$ for C_0, and $Q = +1/6$ for C_i (i=1,2,3), which satisfies the Nishijima-Gell-Mann rule of $Q = I_3 + \frac{B-L}{2}$.

All the leptons and quarks are made of three subquarks as

$$\begin{array}{llll} \nu_e = (w_1 h_1 C_0) & \nu_\mu = (w_1 h_2 C_0) & \nu_\tau = (w_1 h_3 C_0) & \cdots \\ e = (w_2 h_1 C_i) & \mu = (w_2 h_2 C_0) & \tau = (w_2 h_3 C_0) & \cdots \\ u_i = (w_1 h_1 C_i) & c_i = (w_1 h_2 C_i) & t_i = (w_1 h_3 C_i) & \cdots \\ d_i = (w_2 h_1 C_i) & s_i = (w_2 h_2 C_i) & b_i = (w_2 h_3 C_i) & \cdots . \end{array}$$

Furthermore, all the gauge bosons as well as the Higgs scalars are also made of a subquark and an antisubquark as

$$\vec{A}_\mu = (\bar{w}_L \gamma_\mu \tfrac{1}{2} \vec{\tau} w_L)$$

$$B_\mu = (\bar{w}_L \gamma_\mu Y_{w_L} w_L) + (\bar{w}_{1R} \gamma_\mu Y_{w_{1R}} w_{1R}) + (\bar{w}_{2R} \gamma_\mu Y_{w_{2R}} w_{2R}) + \cdots$$

$$\phi = (\bar{w}_{2R} w_L)$$

$$G^a_\mu = (\bar{C}\gamma_\mu \frac{1}{2}\lambda^a C) \qquad (a=1\sim 8)$$

$$H^a_\mu = (\bar{h}\gamma_\mu \frac{1}{2}\Lambda^A h) \qquad (A=1\sim N^2-1), \text{ etc.}$$

It should be noticed that our discovery of the horizontal gauge symmetry is essential in constructing this subquark model.

The dynamics of subquarks should be the one which effectively reproduces quantum flavor dynamics (QFD), the Glashow-Salam-Weinberg gauge theory of $SU(2)_w \times U(1)$,[6] for the electroweak interactions of leptons and quarks and quantum chromodynamics (QCD),[7] the Yang-Mills gauge theory of $SU(3)_C$, for the strong interaction of quarks. A theory in which a gauge theory appears as an effective theory is called "pregauge" theory. Our unified spinor-subquark model[3] of the Nambu-Jona-Lasinio-Bjorken type[8] is such an example. The model not only reproduces the predictions of Georgi and Glashow[9] for the gauge coupling constants and the Weinberg angle in their grand unification gauge theory (GUT),

$$f^2/g^2 = \text{\# of isodoublets / \# of color triplets} = 1/2$$

and

$$\sin^2\theta_W = \Sigma(I_3)^2/\Sigma Q^2 \leqq 3/10,$$

but also predicts the relations between the masses of the weak bosons, the Higgs scalar, and the subquarks,

$$m_W = m_Z\cos\theta_W = \sqrt{3}\left[\frac{m^2_{w_1}+m^2_{w_2}}{2}\right]^{1/2}$$

and

$$m_\eta = 2\left[\frac{m^4_{w_1}+m^4_{w_2}}{m^2_{w_1}+m^2_{w_2}}\right]^{1/2} .$$

These relations indicate that

$$\sqrt{\langle m_w\rangle^2} \cong 46 \text{ GeV} \quad \text{and} \quad m_\eta \geqq (2/\sqrt{3})m_W \cong 92 \text{ GeV}.$$

Since the effective masses of wakems are relatively small in this model, it can be expected that future e^+e^- colliding beam experiments may observe the structures of Z' and $(\bar{w}w)$ continuum states just above the peak of Z.

1. Origin of Flavors and Colors

Most of the subquark models proposed so far can be classified into two categories: the "maximal" and the "minimal"[10] (or the "conservative" and the "radical"[11]). In our type of subquark model,[3] the flavor and color of a lepton or a quark come from the weak-isospin, horizontal, and color quantum numbers of constituent subquarks. Each one of subquarks has only one of these quantum numbers and, therefore, has only one function. Thus, subquarks are what cannot be or need not be further divided. They can be the most fundamental form of matter. In contrast, in subquark models of Harari-Shupe type[12] and the Casalbuoni-Gatto type,[13] the flavors and colors seem to have no definite meaning at the subquark level and to appear as certain combinations of subquarks at the lepton-quark level. it seems to me, however, that the Harari-Shupe subquarks need additional quantum numbers which specify the order of subquarks. If one compares these three models as

whC type	Harari-Shupe type	Casalbuoni-Gatto type
$\nu = (w_1 h C_0)$	$\nu = V_1V_2V_3$	$= \chi$
$\ell = (w_2 h C_0)$	$\bar{d}_i = T_1V_2V_3, V_1T_2V_3, V_1V_2T_3$	$= a_i^+\chi$
$u_i = (w_1 h C_i)$	$u_i = T_1T_2V_3, T_1V_2T_3, V_1T_2T_3$	$= a_j^+a_k^+\chi$
$d_i = (w_2 h C_i)$	$e^+ = T_1T_2T_3$	$= a_1^+a_2^+a_3^+\chi$,

one can find that they are effectively equivalent to each other at the lepton-quark level if the relations of

$$w_1C_0 = V_1V_2V_3 = \chi$$

$$\bar{w}_2\bar{C}_i\bar{w}_1\bar{C}_0 = T_i\bar{V}_i = a_i^+ ,$$

$$w_2\bar{w}_1 = \bar{T}_1\bar{T}_2\bar{T}_3\bar{V}_1\bar{V}_2\bar{V}_3 = b^+ ,$$

and

$$C_0C_1C_2C_3w_1^2w_2^2 = 1$$

hold. A main difference lies in the fact that in our type of models the usual bosons are 2-body states of a subquark and an antisubquark while in models of the Harari-Shupe type they are 4- or 6-body states of subquarks as

$$W^+ = T_1T_2T_3V_1V_2V_3, \quad Z = \bar{T}_1T_1\bar{T}_2T_2\bar{T}_3T_3+\cdots, \quad G = \bar{T}_2V_2\bar{V}_3T_3, \cdots .$$

It seems, therefore, harder to make an effective theory for the lepton and quark interactions in the latter type of models.

Recently, Yasuè has proposed what he calls "subkoma" model,[14] which in a sense combines these two different types of models. In his model, subquarks are further made of two or three subkomas as

$$w_{1L} = (\uparrow\downarrow) \quad w_{1R} = (\uparrow\uparrow) \quad w_{2L} = (\downarrow\uparrow) \quad w_{2R} = (\downarrow\downarrow)$$

$$C_0 = (\uparrow\uparrow\uparrow) \quad C_{1R} = (\downarrow\downarrow\uparrow) \quad C_2 = (\downarrow\uparrow\downarrow) \quad C_3 = (\uparrow\downarrow\downarrow).$$

A very similar model has later been discussed by Mansouri.[15] Also very recently, Maki[10] has proposed another "sub-subquark" model in which the Harari-Shupe's subquark are further made of a scalar Λ and a spinor Θ as

$$T \sim (\Theta\Lambda) \quad \text{and} \quad V \sim (\bar{\Theta}\Lambda).$$

2. Lepton and Quark Mass Spectra

Given a subquark model, what one should do is to explain all the properties of leptons and quarks. To explain the existing mass spectrum of leptons and quarks seems to be the most difficult problem since it is directly related with the unknown dynamics of subquarks. We must imagine what the force between subquarks is and how it binds subquarks into a lepton or a quark. Possible candidates for such a binding force are the following:

1) gravity

Some time ago, Markov proposed an idea of "maximon".[16] Suppose that particles with the mass of order the Planck mass ($G_N^{-1/2} \sim 10^{19}$ GeV) are bound together at the distance of order the Planck length ($G_N^{1/2} \sim 10^{-33}$ cm). The binding energy then becomes of order the Planck mass as

$$G_N G_N^{-1/2} G_N^{-1/2} / G_N^{1/2} \sim G_N^{-1/2},$$

so that the bound state may be very light. That subquarks which may be as heavy as the Planck mass are bound by gravity into a lepton or a quark is a very attractive possibility and can not be discarded.

2) Higgs-exchange force

A similar idea has been proposed by Veltman and Derman.[17] It is the possibility that the force of Higgs-scalar exchange binds subquarks which may be as heavy as the "Fermi mass" ($G_F^{-1/2} \sim 300$ GeV) into a light lepton or quark since

$$G_F G_F^{-1/2} G_F^{-1/2} / G_F^{1/2} \sim G_F^{-1/2}.$$

3) magnetism

Recently, Pati, Salam, and Strathdee[18] have been advocating magnetism as the origin of subquark binding. In their picture, subquarks are Dirac's monopoles or Schwinger's dyons and are bound into a lepton or a quark by the very strong magnetic force due to the large magnetic charges of subquarks given by the Dirac's relation

$$eg/4\pi = n/2 \quad (n = 0, \pm 1, \pm 2, \cdots).$$

4) supersymmetric force

It seems to me also an attractive possibility that Miyazawa's supersymmetry[19] is (at elast approximately) respected by the subquark dynamics.[20] Suppose that leptons and quarks are made of a spinor subquark w_a (a=1,2) and a scalar subquark C_α (α=0,1,2,3). We can then make the supercurrent of subquarks,

$$s_\mu^{a\alpha} = M\gamma_\mu w_a C_\alpha - i\gamma_\nu \gamma_\mu w_a \partial^\nu C_\alpha \ .$$

If we assume PCSC (partially conserved supercurrent hypothesis), the relation

$$\partial^\mu s_\mu^{a\alpha} = F_{a\alpha} m_{a\alpha}^2 f_{a\alpha}$$

holds where f, m, and F are a lepton or quark field, the mass, and the decay constant, respectively. It is very nice to be able to understand qualitatively why leptons and quarks are so light in this picture since they can be taken as Nambu-Goldstone fermions.[21]

5) subcolor force

The above four possibilities certainly deserve closer investigations. The fifth possibility of subcolor force[22] seems to become increasingly popular. Suppose that each subquark has further three subcolor degrees of freedom and that the interaction of subquarks can be described by QSCD (quantum subchromodynamics), the Yang-Mills gauge theory of, say, $SU(3)_{SC}$. It is then natural to expect that subquarks can make leptons, quarks, and bosons and are confined in them, all of which are subcolor-singlet states of subquarks as

$$\ell,\ q \sim \varepsilon_{ijk} s_i s_j s_k$$

and

$$\vec{A},\ B,\ \phi,\ G^a \sim \bar{s}_i s_i \ .$$

This repetition of QCD at the subquark level looks somewhat absurd but seems to be one of the most likely possibilities so that it may deserve closest future investigations. Recently, Weinberg[23] has suggested that color forces may be responsible for at least part of the current-algebraic masses of leptons and quarks. Unfortunately, I have no time to discuss his sofisticated mechanism of producing lepton and quark masses from conspiracy between color and subcolor.

What does the mass formula for leptons and quarks look like if there is any mass formula at all? The quantum numbers possessed by a lepton or a quark include the charge (Q), the weak isospin (I), the baryon and lepton numbers (B and L), and the "generation (or family) number" (n). The lepton and quark masses must be a function of these quantum numbers. A simplest candidate for such mass formula is given by[20]

$$m = m_\mu Q^2 |B-L|^{-3}(n-1)^4 \quad (n=1,2,\cdots).$$

This indicates

$$\begin{bmatrix} m_{\nu_e} & m_{\nu_\mu} & m_{\nu_\tau} & \cdots \\ m_e & m_\mu & m_\tau & \cdots \\ m_u & m_c & m_t & \cdots \\ m_d & m_s & m_b & \cdots \end{bmatrix} \cong m_\mu \begin{bmatrix} 0 & 0 & 0 & \cdots \\ 0 & 1 & 16 & \cdots \\ 0 & 12 & 192 & \cdots \\ 0 & 3 & 48 & \cdots \end{bmatrix},$$

which reproduces the lepton and quark masses remarkably well if $m_t \cong 20$ GeV. Then n^4-dependence of Barut[24] who has predicted the fourth charged lepton at about 9 GeV seems to be ruled out by the recent experiments at PETRA[25] if there is the fourth charged lepton at all. Some other possibility such as a^{n-1}-dependence would be more preferable. Also, the $|B-L|^{-3}$ dependence can be replaced by the $|B-L|^{-2}$ dependence if the mass renormalization effect[26] is taken into account. In any case, to explain the mass spectrum of leptons and quarks seems to be the most important clue to the subquark dynamics though it may be the hardest.

3. Quark (and Lepton) Mixings

There are two ways to understand why the Cabibbo-GIM-KM quark mixings[27] occur for the weak charged current. One is the "hakam mixing", which means that quark mixings are caused by the intrinsic mixings of h_i $(i=1,2,\cdots,N)$. If this is the case, since a h-subquark is shared by a lepton and a quark, the mixing angles for leptons and quarks would become of the same order of magnitude. This possibility would become relevant, related with the recent experimental relevance of possible neutrino oscillation. However, the other possibility, the possibility of "level mixing",[28] seems more natural. A main point of this idea can be traced back to my following suggestion: The weak charged current which has been written in terms of hadrons and quarks can be most fundamentally written in terms of subquarks as

$$J_\mu = \frac{G^\beta}{G^\mu}\bar{p}\gamma_\mu(1 - \frac{g_A^\beta}{g_V^\beta}\gamma_5)n + \frac{G^\Lambda}{G^\mu}\bar{p}\gamma_\mu(1 - \frac{g_A^\Lambda}{g_V^\Lambda}\gamma_5)\Lambda + \cdots$$

$$= U_{ud}\,\bar{u}\gamma_\mu(1-\gamma_5)d + U_{us}\,\bar{u}\gamma_\mu(1-\gamma_5)s + \cdots$$

$$= \bar{w}_1\gamma_\mu(1-\gamma_5)w_2 \ .$$

Let us also suppose that the origin of quark (and lepton) generations (or families) is dynamical. In other words, a quark (or lepton) of a higher generation (u_n or d_n for $n>1$) is considered to be an excited state of its corresponding one of the lowest generation (u or d). Then, the mixing matrix of quarks U_{mn} can be defined by the matrix element of the subquark current between the m-th up-like quark and the n-th down-like quark as

$$\langle u_m | \bar{w}_1\gamma_\mu w_2 | d_n\rangle = U_{mn}\bar{u}_m\gamma_\mu d_n \ .$$

An immediate consequence of this picture is that the Cabibbo angle and angles alike may vary as a function of momentum transfer between quarks.

The algebra of subquark currents[29] includes the familiar commutation relation of isospin currents,

$$\delta(x_0-y_0)\,[V_0^+(x),V_0^-(y)] = 2\delta^4(x-y)V_0^3(x) \ .$$

This relation sandwiched between quark states leads to the (at least approximate) unitarity of the quark mixing matrix, i.e.[30]

$$UU^+ = U^+U = 1$$

if the intermediate states (at least approximately) form a complete set as

$$|u_\ell\rangle\langle u_\ell| = 1 \qquad \text{and} \qquad |d_\ell\rangle\langle d_\ell| = 1 \ .$$

If the isospin breaking is perturbative, the mixing matrix elements are given by

$$U_{mn} = \frac{\langle u_m|H_I|u_n\rangle}{m_{u_m}-m_{u_n}} + \frac{\langle d_m|H_I|d_n\rangle}{m_{d_n}-m_{d_m}} \qquad \text{for } m \neq n.$$

This indicates that the mixing matrix element between different generations decreases as fast as or faster than the inverse of mass difference between the relevant quarks. This property of the quark mixing matrix as well as the above mentioned automatic unitarity of it seems to be very natural and can be taken as a successful consequence of the subquark model. Recently, Akama and I, and independently, Tomozawa[31] have explicitly calculated the quark (and lepton) mixing matrix and its momentum dependence in potential models for the force between subquarks. The results are perfectly consistent with the experimental data. Also, Wu has extensively discussed the neutrino mixing in a subquark model in his recent paper.[32] In short, how to explain the quark mixings in a subquark model has become very transparent.

4. Lepton and Quark Magnetic Moments

Since Glück and Lipkin[33] raised the question why the magnetic moments of leptons and quarks are so close to the Dirac ones if leptons and quarks have the sub-structure, many detailed theoretical analyses have been made.[34] The conclusion is just what is expected: the anomalous magnetic moment (Δa) of a lepton (or quark) due to the sub-structure is, in most cases, proportional either to the inverse of the subquark mass (M) or to the size of the lepton (or quark) (L) as

$$\Delta a \sim \frac{m}{M} \quad \text{or} \quad mL$$

where m is the lepton (or quark) mass. Therefore, there is no problem about the magnetic moment, provided the subquark mass is sufficiently large or the lepton (or quark) size is sufficiently small. Our latest result[35] is given by

$$\Delta a = -2\,\frac{m}{M}\,,$$

which gives a lower bound on the subquark mass of $M > 1.3\times10^{7}$ GeV from the comparison between the experimental and theoretical values for the muon magnetic moment,[36] $|\Delta a| < 2\times10^{-8}$. It should be noticed that a crude estimate of the decay rate of $\mu \to e + \gamma$ and the experimental upper bound give a stronger lower bound on the subquark mass of $M > 1.1\times10^{11}$ GeV as

$$\frac{\Gamma(\mu \to e+\gamma)}{\Gamma(\mu \to e+\bar{\nu}+\nu)} = \frac{96\pi^3\alpha}{G_F^2 m_\mu^4}\,|F_{1,2}|^2 < 10^{-9} \quad (\text{exp.})$$

$$\text{with} \qquad F_{1,2} = -2\sqrt[4]{2}\,(3-2\sqrt{2})\,\frac{m_e+m_\mu}{M}\,.$$

In any case, the problem of the magnetic moment of a lepton (or quark) has turned out to be a source to give a valuable information on the subquark mass and the lepton (or quark) size. A related subject is to explain why the lepton and quark charged currents are so close to the V-A form in spite of their possible sub-structure. I leave this subject for future investigations.

5. Leptons and Quarks of Spin 3/2

In any subquark model in which an ordinary lepton or quark of spin 1/2 is made of subquarks, an abnormal lepton or quark of spin 3/2 or higher can also be made as a different spin configuration of subquarks or as an angular excited state. There is, however, no indication of existence of such an abnormal lepton or quark in GeV region. Therefore, such abnormal leptons and quarks of spin 3/2 or higher, if any, should be much heavier. Why are they much heavier, then? Is it because a spin-dependent force between subquarks is extremely strong? It seems quite certain that the absence of such abnormal leptons and quarks gives some very important information on the subquark dynamics. 't Hooft and Pati[37] have suggested, according to the so-called "Veltman theorem", that abnormal leptons and quarks of spin 3/2 or higher do not exist with the mass much smaller than the inverse of the lepton or quark size because, if otherwise, renormalizability of the effective theory would break down. The theorem, although not proved yet, seems to be something like this: If the mass of a bound state is sufficiently small compared to the inverse size and if perturbation theory can be applied to the force between such bound states, an effective theory which can describe the dynamics of the bound states must be renormalizable. Recently, Weinberg and Witten[38] have proved two theorems: 1) In all theories with a Lorentz-covariant conserved current, there cannot exist composite or elementary massless particles with nonvanishing values of the corresponding charge and with spins higher than 1/2. 2) In all theories with a Lorentz-covariant energy-momentum tensor, composite as well as elementary massless particles with spins higher than 1 are forbidden. They have also claimed that these theorems remove the problem of abnormal leptons and quarks of spin 3/2 or higher. Their assumption that in the absence of electroweak or other perturbations, the lepton and quark masses would vanish seems to leave some open questions. In any case, further investigations are necessary to solve this problem more directly from the subquark dynamics. I must mention here that very recently, Kugo and Uehare[39] made an extension of the theorem of Weinberg and Witten.

6. Exotic States of Subquarks

In any subquark models proposed so far, there may appear many unnecessary (or unknown) fermions and bosons due to exotic combinations of subquarks. It is necessary to explain why all of these exotic states of subquarks have masses much larger than GeV. It is a dynamical problem and, therefore, hard in general. Suppose that the subquark dynamics is described by QSCD. It seems then promising to solve the bound-state problem in a lattice gauge theory[40] of QSCD. It seems no longer necessary to take the lattice as a technical tool. Suppose that the lattice constant a is extremely small, say, of order the Planck length ($G_N^{1/2} \sim 10^{-33}$cm). Then, it does not seem to have any contradiction. It is not necessary to take the limit of $a \to 0$. I would like to consider the lattice as physical. In any case, one can analyze the subquark dynamics in QSCD by making use of all the techniques developed in QCD. In fact, Nussinov[41] has recently insisted the effectiveness of the 1/N expansion in QSCD of $SU(N)_{SC}$. I can think of a lot more applications of the QCD techniques to QSCD problems.[42]

In particular, I would like to emphasize that an application of the Matsumoto's theory of quark confinement in QCD[43] to QSCD seems to be very promising. His theory is based on the suggestion of Banks-Rabinovici, Fradkin-Shenker, and 't Hooft[44] that the confining phase and the Higgs phase are identical. Suppose that leptons and quarks are made of two scalar subquarks s_a^i's (with $s_{ai}^* s_a^j = \delta_{ij}$) and a spinor subquark h^i (a and i are the flavor-color and subcolor indices, respectively). The leptons and quarks, and the gauge bosons are then made as

$$f_{ab} = \varepsilon_{ijk} s_a^i s_b^j h^k \quad \text{and} \quad (A_a{}^b)_\mu = \frac{1}{ie}(D_\mu s_a)^i s_{bi}^* .$$

One can find that these fields obey the ordinary equations of motion in a gauge theory. A similar discussion has recently been made by Bars and Günaydin in their ternary algebraic model and Abbott and Farhi in the Weinberg-Salam model.[45] Also, I must mention that the problem on exotic states of subquarks has been extensively discussed by Casalbuoni, Domokos, and Kovesi-Domokos.[11]

7. Generation (or Family) Problem

Possible candidates for the origin of lepton and quark generation (or family) can be classified into 1) hakams, 2) dynamical excitations, 3) combinations, and 4) others. In 1), there exist <u>a priori</u> hakams, the subquarks with quantum numbers of the horizontal gauge symmetry.[4] In 2), as discussed in 3, leptons and quarks of the second or higher generation are some dynamical (e.g., radial) excitations of the corresponding leptons and quarks of the first generation, which are the ground states. In 3), there are two or more independent states of subquarks which have the same quantum numbers for leptons and quarks.[46] We should test these different possibilities one by one in detail, by finding whether the mass spectrum of leptons and quarks discussed in 2 and the quark mixings discussed in 3 can be explained quantitatively there. Kakazu has given us more details on the third possibility in Ref. 47.

8. Techniquarks vs. Subquarks

For the last several years, Susskind and his collaborators[48] have extensibly investigated the "technicolor hypothesis" that there exist a new family of fermions called "techniquarks", (U_i, D_i) ($i=1\sim N$), which form a doublet of $SU(2)_w$ and a N-plet of a new gauge symmetry of $SU(N)_{TC}$. They claim that the weak interaction symmetry breaking scale ($\sim$300 GeV) can naturally arise as a consequence of a strongly interacting "technicolor" gauge theory at a scale of order 1 TeV. Their techniquarks U and D with technicolor play the same role as our subquarks w_1 and w_2 with subcolor in producing composite Higgs scalars dynamically. Therefore, it seems that there must be much in common in techniquark and subquark models as far as phenomenology in the weak interactions at low energies is concerned if the critical energy scale of subcolor is as low as 1 TeV. However, I would like to emphasize the trivial and yet essential difference that techniquarks are "brothers (or sisters)" of leptons and quarks while subquarks are constituents of them. It seems very instructive to remember the difference between the Sakata model with (p, n, Λ) and the quark model of Gell-Mann and Zweig with (u, d, s). We should make a list of phenomenological differences between the techniquark model and the subquark model so that future experiments may clearly tell us which picture nature favors.

9. Grand Unification vs. Subquarks

In a grand unified gauge theory (GUT),[49] the grand unification of all the lepton and quark interactions is expected to occur at energy of order 10^{15} GeV where the strong and electroweak interactions would have an equal strength. If the mass scale of subquarks (or the energy scale of subquark dynamics) is larger than 10^{15} GeV, subquarks (or the sub-structure of leptons and quarks) will appear after the grand unification. In this case, both grand unification and subquarks will meet a peaceful co-existence. Moreover, the case where the mass scale of subquarks is of order the Planck mass or even larger is the most preferable since subquarks would then appear after the supergrand unification of the strong and electroweak interactions with gravity. If it is below 10^{15} GeV, the idea of grand unification seems to be an illusion. Even in this case, however, it is yet possible that some recent results calculated in grand unification theories (GUT's) for the proton decay, etc.,[26] may remain to be approximately valid if GUT can be taken as an effective theory. The comparison between GUT's and subquark models reminds me of that between "bootstrap" theories of hadrons and quark models of them. Again, the future will tell us which will win the game (or both will win?).

10. Subquark Supergravity vs. Subquark Pregeometry

As is well known,[50] the SU(8) supergravity as a best candidate for the supergrand unification theory (GUT) has a difficulty that the multiplets of spin 1/2 and spin 1 contained in the theory are too small to accommodate the known fundamental fermions and bosons such as μ and $W^{\pm}$. Salam[51] has suggested that the difficulty may be overcome if such fermions and bosons that can not be accommodated may be taken as composites of subquarks. Then, Curtwright and Freund[52] have proposed the extended supergravity model of SU(8) in which the fundamental octet contains a doublet of wakems, a triplet of hakams, and a triplet of chroms as

$$\underline{8} = (w_1\ w_2\ h_1\ h_2\ C_1\ C_2\ C_3).$$

If the SU(8) model works, it will indicate the triumph of "subquark supergravity", the supergrand unification in terms of subquarks.

In the above subquark supergravity, the graviton is elementary although gauge bosons as well as leptons and quarks are composite of subquarks. Akama and I[53] have proposed a model of "subquark pregeometry" in which the graviton is also a composite of a subquark-antisubquark pair. The compositeness of the graviton can be traced back to Sakharov in 1967 in his pregeometry[54] that the Einstein's gravity is a quantum effect of matter. The graviton is taken as a collective excitation of matter fields. If the gauge bosons and Higgs scalars as well as leptons and quarks are all composites of subquarks as in our subquark model, why not is the graviton also a composite of them. Everything is made of subquarks and every force is due to them. But, is it possible? The answer is affirmative. In fact, we have demonstrated in a simple subquark model of pregauge and pregeometry that the Yang-Mills gauge theory and the Einstein's general relativity for leptons and quarks can be effectively reproduced by the subquark dynamics as follows:

$$L = \bar{w}[e^{k\mu}\gamma_k(iD_\mu+\lambda^a A^a_\mu)-M]w + g^{\mu\nu}\partial_\mu C^+\partial_\nu C - \mu^2 C^+ C$$

$$+ \bar{f}wC + C^+\bar{w}f + c_0 - c_1 g^{\mu\nu}A^a_\mu A^a_\nu - c_2\bar{f}f$$

$$\rightarrow L_{eff} = \frac{1}{16\pi G}R - \frac{1}{4}g^{\mu\nu}g^{\rho\sigma}F^a_{\mu\rho}F^a_{\nu\rho}$$

$$+ \bar{f}[e^{k\mu}\gamma_k(iD_\mu + \frac{\lambda^a}{2}A^a_\mu) - m]f + \cdots$$

$$\text{with} \quad F^a_{\mu\nu} = \partial_\mu A^a_\nu - \partial_\nu A^a_\mu + ef^{abc}A^b_\mu A^c_\nu .$$

This model indicates that there must exist the short-distance cutoff at about the Planck length,

$$\Lambda^{-1} \sim G^{1/2} ,$$

and that the Newtonian gravitational constant G and the gauge coupling constant e are related with each other as

$$e^2 \sim 1/\ell n\ (1/G).$$

Both of these confirm the old conjecture made by Landau in 1955.[55] The model also suggests that the photon may have an extremely small but non-vanishing mass. The present experimental upper bound on the photon mass[56] is 6×10^{-25} GeV. How to improve the upper bound or to detect the possible minute photon mass is an important future problem.

11. Conclusion

In conclusion, as I have discussed in this chapter, a few of the fundamental problems in subquark models have been solved, but most of them still remain to be solved. I believe that subquarks are there and working. Much more efforts would be needed before finding the true theory of subquarks, which may be the final theory in physics.

References and Footnotes

1. For a partial list of references, see for example H. Terazawa, Phys. Rev. D22, 184 (1980). For a review of the history of related works, see especially the Section II. Some of the earlier references include C.-K. Chang, Phys. Rev. D5, 950 (1972); J. C. Pati and A. Salam, Phys. Rev. D10, 275 (1974); K. Matumoto, Prog. Theor. Phys. 52, 1973 (1974); O. W. Greenberg and C. A. Nelson, Phys. Rev. D10, 2567 (1974); O. W. Greenberg, Phys. Rev. Lett. 35, 1120 (1975); J. C. Pati, A. Salam, and J. Strathdee, Phys. Lett. 59B, 165 (1975).
2. H. Terazawa, in Proc. 1981 INS Symp. on Quark and Lepton Physics, edited by K. Fujikawa et al. (INS, Univ. of Tokyo, Tokyo, 1981), p.296.
3. H. Terazawa, Y. Chikashige, and K. Akama, Phys. Rev. D15, 480 (1977).
4. The horizontal gauge symmetry of leptons and quarks was first proposed by us in 1976. See K. Akama and H. Terazawa, INS-Report-257 (INS, Univ. of Tokyo) April 15, 1976 and H. Terazawa, Prog. Theor. Phys. 58, 1276 (1977). For later studies, see, for example, T. Maehara and T. Yanagida, Prog. Theor. Phys. 60, 822 (1978) and F. Wilczek and A. Zee, Phys. Rev. Lett. 42, 421 (1979).
5. Pati, Salam, and Strathdee call wakems, hakams, and chroms "flavons", "familons", and "chromons", respectively. They also call subquarks, in general, "preons". See for example,

A. Salam, Rev. Mod. Phys. 52, 525 (1980).

6. S. L. Glashow, Nucl. Phys. 22, 579 (1961); S. Weinberg, Phys. Rev. Lett. 19, 1264 (1967); A. Salam, in Elementary Particle Physics, edited by N. Svartholm (Almqvist and Wiksell, Stockholm, 1968), p.367.
7. Y. Nambu, in Preludés in Theoretical Physics, edited by A. de Shalit (North-Holland, Amsterdam, 1966); H. Fritzsch and A. Gell-Mann, in Proc. XVI Int. Conf. on High Energy Phys., edited by A. Roberts (NAL, Batavia, 1973), Vol.2, p.135.
8. Y. Nambu and G. Jona-Lasinio, Phys. Rev. 122, 345 (1961); J. D. Bjorken, Ann. Phys. (N.Y.) 24, 174 (1963).
9. H. Georgi and S. L. Glashow, Phys. Rev. Lett. 32, 438 (1974); H. Georgi, H. R. Quinn, and S. Weinberg, Phys. Rev. Lett. 33, 451 (1974).
10. Z. Maki, Soryushiron Kenkyu (Kyoto) 62, F57 (1981) (in Japanese) and also in Proc. 1981 INS Symposium on Quark and Lepton Physics, Tokyo, June 25-27, 1981, edited by K. Fujikawa et al. (INS, Univ. of Tokyo, Tokyo, 1981), p.325.
11. See for example, R. Casalbuoni, G. Domokos, and S. Kovesi-Domokos, Phys. Rev. D23, 462 (1981).
12. H. Harari, Phys. Lett. 86B, 83 (1979); M. A. Shupe, Phys. Lett. 86B, 87 (1979).
13. R. Casalbuoni and R. Gatto, Phys. Lett. 88B, 306 (1979).
14. M. Yasuè, Phys. Lett. 91B, 85 (1980), Prog. Theor. Phys. 65, 1995 (1981), and also in Proc. 1981 INS Symposium on Quark and Lepton Physics, Tokyo June 25-27, 1981, edited by K. Fujikawa et al. (INS, Univ. of Tokyo, Tokyo, 1981), p.332.
15. F. Mansouri, Phys. Lett. 100B, 25 (1981).
16. M. A. Markov, J. Exptl. Theoret. Phys. (U.S.S.R.) 51, 878 (1966) Sov. Phys. JETP 24, 584 (1967) .
17. M. Veltman, Acta Phys. Polon B12, 437 (1981); E. Derman, Phys. Lett. 95B, 369 (1980).
18. J. C. Pati, Phys. Lett. 98B, 40 (1981); J. C. Pati, A. Salam, and J. Strathdee, Nucl. Phys. B185, 416 (1981).
19. H. Miyazawa, Prog. Theor. Phys. 36, 1266 (1966); J. Wess and B. Zumino, Nucl. Phys. B70, 39 (1974); A. Salam and J. Strathdee, Nucl. Phys. B76, 477 (1974).
20. H. Terazawa, Prog. Theor. Phys. 64, 1763 (1980). See also H. Terazawa, Prog. Theor. Phys. 64, 1388 (1980) and Z. Maki, in Ref. 10.
21. D. V. Volkov and V. P. Akulov, Phys. Lett. 46B, 109 (1973).
22. G. 't Hooft, in Recent Developments in Gauge Theories, edited by G. 't Hooft et al. (Plenum Pub. Co., New York, 1980), p.135; R. Casalbuoni and R. Gatto, Phys. Lett. 93B, 47 (1980); H. Terazawa, in Ref. 20; O. W. Greenberg and J. Sucher, Phys. Lett. 99B, 339 (1981). See also K. Matumoto, in Ref. 1.
23. S. Weinberg, in Proc. Workshop on Weak Interactions as Probes of Unification, VPI, Dec. 4-6, 1980, edited by G. B. Collins et al. (AIP, New York, 1981), p.521.
24. A. O. Barut, Phys. Rev. Lett. 42, 1251 (1979).
25. See for example, D. P. Barber et al., Phys. Rev. Lett. 45, 1904 (1980).

26. See for example, A. J. Buras, J. Ellis, M. K. Gaillard, and D. V. Nanopoulos, Nucl. Phys. B135, 66 (1978).
27. S. L. Glashow, J. Iliopoulos, and L. Maiani, Phys. Rev. D2, 1285 (1970); M. Kobayashi and T. Maskawa, Prog. Theor. Phys. 49, 652 (1973).
28. H. Terazawa, Prog. Theor. Phys. 58, 1276 (1977).
29. H. Terazawa, in Ref. 1.
30. H. Terazawa, in Ref. 20; O. W. Greenberg and J. Sucher, in Ref. 22. See also V. Visnjić-Triantafillow, FERMILAB-Pub.-80/34-THY (Fermilab) March, 1980.
31. H. Terazawa and K. Akama, Phys. Lett. 101B, 190 (1981); Y. Tomozawa, UM HE 81-3 (Univ. of Michigan) 1981, in Proc. 1981 INS Symposium on Quark and Lepton Physics, Tokyo, June 25-27, 1981, edited by K. Fujikawa et al. (INS, Univ. of Tokyo, Tokyo, 1981), p.319, and Phys. Lett. 104B, 136 (1981).
32. D. Wu, Phys. Rev. D23, 2038 (1981).
33. M. Glück, Phys. Lett. 87B, 247 (1979); J. Lipkin, Phys Lett. 89B, 358 (1980).
34. K. Akama, Prog. Theor. Phys. 64, 1494 (1980); G. L. Shaw, D. Silverman, and R. Slansky, Phys. Lett. 94B, 57 (1980); P. F. Smith and J. D. Lewin, Phys. Lett. 94B, 484 (1980); S. J. Brodsky and S. D. Drell, Phys. Rev. D22, 2236 (1980); H. Terazawa and K. Akama, Phys. Lett. 96B, 276 (1980); O. W. Greenberg and J. Sucher, in Ref. 22.
35. H. Terazawa, M. Yasuè, K. Akama, and M. Hayashi, Phys. Lett. 112B, 387 (1982).
36. See for example, T. Kinoshita, in Proc. 19th Int. Conf. High Energy Physics, edited by S. Homma, M. Kawaguchi and H. Miyazawa (Phys. Soc. Japan, Tokyo, 1979), p.571.
37. G. 't Hooft, in Ref. 22; J. C. Pati, in Ref. 18.
38. S. Weinberg and E. Witten, Phys. Lett. 96B, 59 (1980).
39. T. Kugo and S. Uehara, KUNS 579 (Kyoto Univ.) April, 1981 and T. Kugo, in Proc. 1981 INS Symposium on Quark and Lepton Physics, Tokyo, June 25-27, 1981, edited by K. Fujikawa et al. (INS, Univ. of Tokyo, Tokyo, 1981), p.336 and Phys. Lett. 109B, 205 (1982).
40. K. G. Wilson, Phys. Rev. D10, 2445 (1974).
41. S. Nussinov, Phys. Rev. Lett. 45, 1912 (1980).
42. For example, applying the approach of Shifman et al. to QCD to QSCD immediately leads to the result that the gauge coupling constant of the lepton and quark interactions is given by $g^2/4\pi \simeq 2\pi/eN$ where N is the number of multiplets. See M. A. Shifman, A. I. Vainstein, and V. I. Zakharov, Nucl. Phys. B147, 385 (1979). Also, the recent calculations of the glue-ball mass in lattice gauge theories can be converted into those of the subglue-ball mass in subcolor lattice gauge theories. We can expect that the subglue-ball mass is as large as $3\sqrt{T}$ (> 1 TeV ?) where T is the sub-string tension. See for example, G. Bhanot, Phys. Lett. 101B, 95 (1981) and N. Kimura, in Proc. 1981 INS Symposium on Quark and Lepton Physics, Tokyo, June 25-27, 1981, edited by K. Fujikawa et al.

(INS, Univ. of Tokyo, Tokyo, 1981), p.108.
43. T. Matsumoto, Phys. Lett. 97B, 131 (1980) and also in Proc. 1981 INS Symposium on Quark and Lepton Physics, Tokyo, June 25-27, 1981, edited by K. Fujikawa et al. (INS, Univ. of Tokyo, Tokyo, 1981), p.128.
44. T. Banks and E. Rabinovici, Nucl. Phys. B160, 349 (1979); E. Fradkin and S. H. Shenker, Phys. Rev. D19, 3682 (1979); G. 't Hooft, in Ref. 22; S. Dimopoulos, S. Raby, and L. Susskind, Nucl. Phys. B173, 208 (1980).
45. I. Bars and M. Günaydin, Phys. Rev. D22, 1403 (1980); L. F. Abbott and E. Farhi, Phys. Lett. 101B, 69 (1981).
46. R. Casalbuoni, G. Domokos, and S. Karesi-Domokos, in Ref. 11; K. Matumoto and K. Kakazu, Prog. Theor. Phys. 65, 390 (1981); M. Yasuè, the second paper in Ref. 14; S. Weinberg, in Ref. 23.
47. K. Kakazu, in Proc. 1981 INS Symposium on Quark and Lepton Physics, Tokyo, June 25-27, 1981, edited by K. Fujiikawa et al. (INS, Univ. of Tokyo, Tokyo, 1981), p.329.
48. L. Susskind, Phys. Rev. D20, 2619 (1979). See also S. Weinberg, Phys. Rev. D13, 974 (1976) and D19, 1277 (1979).
49. J. C. Pati and A. Salam, in Ref. 1; H. Georgi and S. L. Glashow, in Ref. 9.
50. M. Gell-Mann, talk presented at the Washington Meeting of the American Physical Society, April, 1977.
51. A. Salam, in Proc. 19th Int. Conf. on High Energy Physics, Tokyo, edited by S. Homma, M. Kawaguchi, and H. Miyazawa (Phys. Soc. Japan, Tokyo, 1979), p.617.
52. T. C. Curtwright and P. G. O. Freund, Supergravity, edited by P. Van Nieuwenhuizen and D. Z. Freedman (North-Holland, Amsterdam, 1979), p.197; J. Ellis, M. K. Gaillard, and B. Zumino, Phys. Lett. 94B, 343 (1980); J. Ellis, M. K. Gaillard, L. Maiani, and B. Zumino, LAPP-TH-15/Ref.TH-2841-CERN (LAPP) April 1, 1980.
53. H. Terazawa and K. Akama, in Ref. 34 and Phys. Lett. 97B, 81 (1980).
54. A. Sakharov, Doklady Acad. Nauk SSSR 177, 70 (1967); K. Akama, Y. Chikashige, T. Matsuki, H. Terazawa, Prog. Theor. Phys. 60, 868 (1978); S. L. Adler, Phys. Rev. Lett. 44, 1567 (1980).
55. L. Landau, in Niels Bohr and the Development of Physics, edited by W. Pauli (McGraw-Hill, New York, 1955), p.52; Ya. B. Zel'dovich, ZhETF Pis'ma 6, 922 (1967); H. Terazawa, Y. Chikashige, K. Akama, and T. Matsuki, Phys. Rev. D15, 1181 (1977).
56. L. Davis, Jr., A. S. Goldhaber, and M. M. Nieto, Phys. Rev. Lett. 21, 1402 (1975). See also Y. Yamaguchi, Prog. Theor. Phys. Suppl. 11, 33 (1959), which contains discussions on a variety of exotic possibilities ranging from proton decay to non-vanishing photon mass due to charge non-conservation. A possibility of the curvature-dependent photon mass is discussed, relating to the Dirac's large number hypothesis, in H. Terazawa, Phys. Lett. 101B, 43 (1981).

Excercises

1) Construct a subquark model which has only less than 6 independent subquarks of which all quarks and leptons can be made.
2) Suppose that a hakam (h) and chroms (C_α,α=0,1,2,3) make scalar bound states of "di-subquarks" (Call them chroms and let them be C_α also), and describe not only all quarks and leptons but also gauge bosons and Higgs scalars in terms of wakems (w_i,i=1,2) and these scalar chroms (C_α).
3) Invent a supergrand unified subquark model in which all fundamental fermions (quarks and leptons), gauge bosons (the photon, weak bosons, and gluons), Higgs scalars, and also vierbeins or space-time metrices are composite of subquarks.

II. UNIFIED AND GRAND UNIFIED COMPOSITE MODELS

In this chapter, I shall discuss some of the current topics in unified and grand unified composite models in which quarks and leptons are made of subquarks, the more fundamental particles.[1] They include the following subjects:

1. Minimal Model
2. Nucleon Decays
3. Mass Spectrum of Quarks and Leptons
4. Mass Scale for the Sub-Structure
5. Quarks and Leptons as Nambu-Goldstone Fermions
6. 't Hooft Condition
7. "Duality"
8. Flavor Mixing
9. Excited Quarks, Leptons and Gauge Bosons
10. Half-Charged Particles
11. Cosmology with Composite Quarks and Leptons
12. Conclusion

1. Minimal Model

Let us start with introducing the minimal composite model of quarks and leptons as a standard of reference for discussions in this review. It consists of an isodoublet of spinor subquarks w = (w_1,w_2)[2] (called wakems abbreviating weak and electromagnetic) and a color quartet of scalar subquarks C = (C_0,C_1,C_2,C_3)[3] (called chroms meaning color). The charges of these subquarks are Q=+1/2 for w_1, Q=-1/2 fow w_2 and C_0, and Q=+1/6 for C_i (i=1,2,3), satisfying the Nishijima-Gell-Mann rule of $Q = I_3 + \frac{B-L}{2}$ where I_3 is the third component of isospin (I_3=+1/2 for w_1 and I_3=-1/2 for w_2), and B and L are the baryon number and the lepton number, respectively (B=+1/3 for C_i and L=+1 for C_0). The quarks and leptons of the first generation can be taken as composite states of these subquarks as

$$\nu_e = w_1 C_0 \qquad u_i = w_1 C_i$$
$$e = w_2 C_0 \qquad d_i = w_2 C_i \ .$$

The gauge bosons as well as the Higgs scalars can also be taken as composite states of subquark-antisubquark pairs as

$$W_\mu^+ = \bar{w}_{2L} \gamma_\mu w_{1L} \qquad W_\mu^- = \bar{w}_{1L} \gamma_\mu w_{2L}$$

$$A_\mu = \frac{\sqrt{3}}{2}(\frac{1}{2}\bar{w}_1 \gamma_\mu w_1 - \frac{1}{2}\bar{w}_2 \gamma_\mu w_2 - \frac{1}{2} i C_0^+ \overleftrightarrow{\partial}_\mu C_0 + \frac{1}{6} i C_i^+ \overleftrightarrow{\partial}_\mu C_i)$$

$$Z_\mu = \frac{\sqrt{5}}{2}(\frac{1}{2}\bar{w}_{1L} \gamma_\mu w_{1L} - \frac{1}{2}\bar{w}_{2L} \gamma_\mu w_{2L})$$
$$- \frac{3\sqrt{5}}{10}(\frac{1}{2}\bar{w}_{1R} \gamma_\mu w_{1R} - \frac{1}{2}\bar{w}_{2R} \gamma_\mu w_{2R} - \frac{1}{2} i C_0^+ \overleftrightarrow{\partial}_\mu C_0 + \frac{1}{6} i C_i^+ \overleftrightarrow{\partial}_\mu C_i)$$

$$G_\mu^a = \sqrt{2} i C \overleftrightarrow{\partial}_\mu \frac{\lambda^a}{2} C \qquad \text{for } a = 1-8$$

$$\phi = \bar{w}_R w_L \qquad \text{etc.}$$

This minimal composite model may reproduce QFD (the unified gauge theory of Glashow-Salam-Weinberg $SU(2)_w \times U(1)$) for the electroweak interactions of quarks and leptons and QCD (the gauge theory of color $SU(3)_C$) for the strong interaction of quarks as effective theories at low energies. It may also reproduce some results of the grand unified SU(5) gauge theory of Georgi and Glashow,[4] including

$$\sin^2 \theta_w = tr(I_3)^2/trQ^2 = 3/8$$

and

$$f^2/g^2 = tr(I_3)^2/tr(\lambda^a/2)^2 = 1$$

where θ_w is the weak mixing angle and f and g are the $SU(3)_C$ gauge coupling constant and the $SU(2)_w$ one, respectively.

Although the subquark dynamics is totally unknown, it can be assumed to be "quantum subchromodynamics (QSCD)" (the Yang-Mills gauge theory) of subcolor $SU(N)_{SC}$ in which w and C are N-plet and anti-N-plet of subcolor $SU(N)_{SC}$, respectively. Then, the model has the global $SU(2)_w \times U(1)_w \times SU(4)_C \times U(1)_C$ symmetry as well as the local subcolor $SU(N)_{SC}$ symmetry. It naturally produces subcolor singlet states including the composite fermions of (wC) which behave as $(\underline{2},\underline{4})$ in $SU(2)_w \times SU(4)_C$ and the composite bosons of $(\bar{w}w)$ and (C^+C) which behave as $(\underline{3},\underline{1})+(\underline{1},\underline{1})$ and $(\underline{1},\underline{15})+(\underline{1},\underline{1})$, respectively. The best candidate for subcolor symmetry is $SU(4)_{SC}$. If this is the case, it produces additional subcolor singlet states including the composite fermions of $(wwwC^+)$ and $(wC^+C^+C^+)$ which behave as $(\underline{4},\underline{4}^*)+2(\underline{2},\underline{4}^*)$ and $(\underline{2},\underline{4})$, respectively, and the composite bosons of (wwww), (wwC^+C^+) and (CCCC) which behave as $(\underline{5},\underline{1})+3(\underline{3},\underline{1})+2(\underline{1},\underline{1})$, $(\underline{3},\underline{6}^*)+(\underline{1},\underline{6}^*)$ and $(\underline{1},\underline{1})$, respectively. These additional composite fermions which behave as $(\underline{2},\underline{4})$ can be taken as quarks and leptons of a higher generation, which is one of the possibilities for the origin of generations (I shall discuss this and other possibilities in more details later). The model can, therefore, correctly accommodate not only three or four generations of composite quarks and leptons, (wC), $(\bar{w}\bar{w}\bar{w}C)$ and $(wC^+C^+C^+)$, depending on whether $2(\underline{2},\underline{4})$ of $(\bar{w}\bar{w}\bar{w}C)$ are degenerate or not, but also composite gauge bosons, $(\bar{w}w)$ and (C^+C), including the electroweak vector bosons, γ, $W^{\pm}$ and Z, and the gluons, G^a (a=1-8).

2. Nucleon Decays

It is well known that the prediction of proton decay with the life time of

$$\tau(p \to e^+\pi^0) \cong 2 \times 10^{29} y \left(\frac{m_X}{4\times 10^{14}\,\mathrm{GeV}}\right)^4$$

in the original Georgi-Glashow grand unification model of SU(5)[4] has been jeopardized by the recent Irvine-Michigan-Brookhaven data for proton decays.[5]

It is also one of the serious difficulties in composite models of the Harari-Shupe type[6] that baryon-number changing nucleon decays occur too fast due to a simple exchange of subquarks. In composite models of our type, on the other hand, baryon-number seems to be conserved since the chroms have definite baryon or lepton numbers, i.e., B=0 and L=1 for C_0 and B = +1/3 and L=0 for C_i (i=1,2,3). This is, however, not true. Due to the possible condensation of subcolor-singlet $(C_0C_1C_2C_3)$, which may happen in the minimal model, baryon-number conservation can possibility break down, causing baryon-number changing nucleon decays. In this picture of the baryon-number violation, B-L conserves as in the grand unification gauge model. Also, it is remarkable that generation-changing nucleon decay modes such as $p \to K^0 e^+$, $K^+ \bar{\nu}_e$, $\mu^+ \gamma$, $\mu^+ \gamma\gamma$ etc. may be enhanced compared to generation-conserving ones such as $p \to e^+ \pi^0$, which is the most enhanced in the grand unification gauge model. This is because the former modes are due to the condensation of $(C_0 C_1 C_2 C_3)$ while the latter are due to that of $(C_0 C_1 C_2 C_3)$ and $(w_1 w_1 w_2 w_2)$ (See the details in Fig. 1). Whether

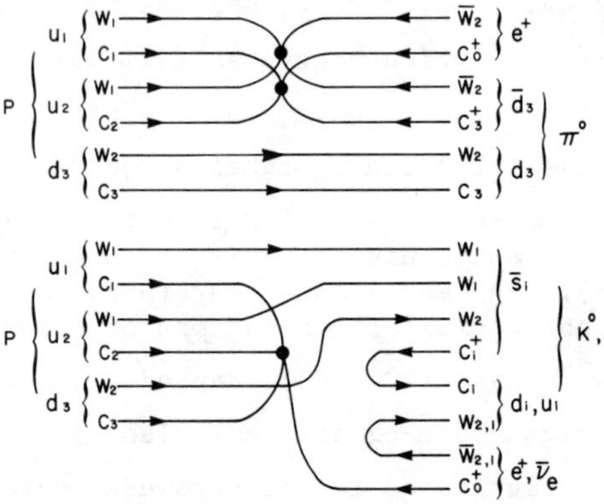

FIG.1. The subquark diagrams for proton decay.

this picture of nucleon decays is working or not can be checked by the KAMIOKANDE experiment which is under way.[5]

3. Mass Spectrum of Quarks and Lepton

One of the most difficult problems to be solved in composite models is to explain the existing mass spectrum of quarks and leptons:

$$m\begin{pmatrix} \nu_e & \nu_\mu & \nu_\tau & \cdots \\ e & \mu & \tau & \cdots \\ u & c & t & \cdots \\ d & s & b & \cdots \end{pmatrix} = \begin{pmatrix} \lesssim 10\text{-}20\text{eV} & <.52\text{MeV} & <.164\text{GeV} & \cdots \\ \begin{matrix}.5110034\\(14)\text{MeV}\end{matrix} & \begin{matrix}105.65943\\(18)\text{MeV}\end{matrix} & \begin{matrix}1784.2\\(3.2)\text{MeV}\end{matrix} & \cdots \\ \sim 4\text{-}5\text{MeV} & \sim 1.2\text{GeV} & \gtrsim 20\text{GeV} & \\ \sim 8\text{-}9\text{MeV} & \sim .15\text{GeV} & \sim 5\text{GeV} & \cdots \end{pmatrix} .$$

The gross feature of the mass spectrum can be taken as either one of

$$m_\mu \begin{pmatrix} 0 & 0 & 0 & \cdots \\ 0 & 1 & \sim 16.9 & \cdots \\ 0 & \sim 11 & \gtrsim 190 & \cdots \\ 0 & \sim 1 & \sim 47 & \cdots \end{pmatrix}, \quad m_\tau \begin{pmatrix} 0 & 0 & 0 & \cdots \\ 0 & 0 & 1 & \cdots \\ 0 & \sim 1 & \gtrsim 11 & \cdots \\ 0 & 0 & \sim 2.8 & \cdots \end{pmatrix}$$

$$\text{or}\quad m_L \begin{pmatrix} 0 & 0 & 0 & 0\cdots \\ 0 & 0 & 0 & 1\cdots \\ 0 & 0 & \sim 1? & \gtrsim 11?\cdots \\ 0 & 0 & 0 & \sim 3?\cdots \end{pmatrix}.$$

There seems to be no simple empirical mass formula except for[7]

$$m_{q,\ell} = m_\mu \frac{Q^2}{|B-L|^3}(n-1)^4 = m_\mu \begin{pmatrix} 0 & 0 & 0 & 0\cdots \\ 0 & 1 & 16 & 81\cdots \\ 0 & 12 & 192 & 972\cdots \\ 0 & 3 & 48 & 243\cdots \end{pmatrix}$$

where Q, B, L and n are the charge, baryon number, lepton number and generation number of quarks and leptons, respectively. Since the possibility of the fourth charged lepton with the mass of order $81m_\mu$ has already been excluded by the e^+e^- colliding beam experiments,[8] the $(n-1)^4$ dependence should be modified if there exists more than three generations of quarks and leptons. The exponential dependence of $m_\mu a^{n-1}$ would be better in that case although it does not work well for lower generations. In any case, how to justify this type of mass formulas is a problem to be solved. I shall present an answer to this question later. It is often emphasized that the masses of the fundamental fermions (m_f), quarks and leptons, are much smaller than their size inverses (Λ_f) although the definition of the size of quarks and leptons is not clear. What seems to be more definite is that the masses of the fundamental fermions are much smaller than the masses of the weak bosons $W^\pm$ and Z:

$$m_f \ll m_{W^\pm},\ m_Z < \Lambda_f\ ?$$

However, this may not be the case either if the top-quark mass is much larger than usually expected. In fact, our sum rule[2]

$$\sqrt{\langle m_f^2\rangle} = m_{W^\pm}/\sqrt{3} \cong 46\text{GeV}$$

indicates[9] $m_t \cong 131$GeV if there exist only three generations of quarks and leptons. This seems to be excluded by the latest analysis of the UA1 data[10] in which they claim to have found the top quark with the mass lying between 30 and 50 GeV. However, the sum rule also indicates that the average mass of quarks and leptons of the fourth generation is of order $m_{W^\pm}$ and m_Z, i.e.,

$$\sqrt{\langle m_f^2 \rangle_{n=4}} \cong 2m_{W^\pm}/\sqrt{3} \cong 92\text{GeV} ,$$

if there exist four generations of quarks and leptons. Our sum rule which indicates that the average mass of quarks and leptons is of the same order as the masses of the weak bosons seems to be natural in composite models where both the fundamental fermions and the gauge bosons are composites of subquarks although it has originally been derived in a particular model. In any case, the most precise statement is that the masses of quarks and leptons of lower (at least the first and second) generations are much smaller than the masses of the weak bosons. But, why? I shall give an answer to this question later.

4. Mass Scale for the Sub-Structure

What is the mass scale for the sub-structure? Where does it come from? These are also difficult questions to be answered at this stage. As it stands now, the mass scale can lie between several 10GeV and 10^{18}GeV. Very lately Yamada has given us a review on the experimental constraints on the mass scale in more detail in Ref. 11.

The smallest possible mass scale comes from the weak boson mass. The most naive expectation is that the mass of the subquark w is roughly a half of those of the weak bosons $W^\pm$ and Z since the latter are made of a subquark and an antisubquark, provided that the subquarks are loosely bound. More precisely, suppose that the effective Yukawa coupling constant (G_Y) of the composite Higgs scalar (ϕ) and the subquark (w) is of the same order as the effective gauge coupling constant (g) of the composite gauge bosons ($W^\pm$ and Z), i.e. $G_Y \sim g$. Then, it is natural to expect that the subquark mass is of the same order as the weak boson mass, i.e. $m_w \sim m_{W^\pm}$, m_Z, provided that both of these masses are generated by the vacuum expectation value of the Higgs scalar, i.e. $m_w \sim G_Y\langle\phi\rangle_0$ and $m_{W^\pm}$, $m_Z \sim g\langle\phi\rangle_0$. In fact, our sum rule in a dynamical model[2,12] states

$$\sqrt{\langle m_w^2 \rangle} = m_{W^\pm}/\sqrt{3} \cong 46\text{GeV}.$$

If indeed the mass scale for the sub-structure is so small, it can be seen not only in future experiments at LEP but possibly in experiments at SPS $\bar{p}p$ Collider. The best way to find such sub-structure is to search for the excited gauge bosons ($W^{\pm\prime}$ and Z')[12] at their masses of order 100-200GeV and for the excited quarks and leptons (u^*, d^*, e^*, μ^*, etc.)[13] at their masses of order 50-100GeV.

The lower bounds on the mass scale for the sub-structure of order 100GeV-1TeV have been obtained by e^+e^- colliding beam experiments. The conventional parametrization of the sub-structure by modifying the photon propagator by a factor of $1\pm(q^2/\Lambda^2)$ and the data for $e^+e^- \to e^+e^-$, $\mu^+\mu^-$, $\tau^+\tau^-$ have given the result of $\Lambda \gtrsim$ 100-200GeV. A new parametrization of Eichten, Lane and Peskin[14] with the residual interaction of subquark dynamics of the type

$$L = (g^2/2\Lambda^2)[\eta_{LL}\bar{\psi}_L\gamma_\mu\psi_L\bar{\psi}_L\gamma^\mu\psi_L + \cdots\cdots] ,$$

where the constants $g^2/4\pi$ and η_{LL} are assumed to be of order unity, and the same experimental data have given the result of $\Lambda \gtrsim$ 750-1500GeV. It is clear that the magnitude of the lower bound strongly depends on the parametrization of the mass scale for the sub-structure.

A more stringent lower bound of order 10^3-10^4TeV has been obtained from comparison between experimental and theoretical values for the electron and muon anomalous magnetic moments. In most of the model calculations,[15] the effect of the sub-structure on the lepton anomalous moment turns out to be

$$\Delta a_\ell \sim m_\ell/\Lambda$$

where Λ is the size inverse of the lepton or the mass scale of the subquarks of which the lepton is made. On the other hand, the most precise comparison between experimental and theoretical values[16] reports

$$a_\ell^{\text{exp.}} - a_\ell^{\text{theor.}} = \begin{cases} -(2.51\pm1.54)\times10^{-10} & \text{for } e \\ (0.38\pm1.1)\times10^{-8} & \text{for } \mu . \end{cases}$$

By comparing this with the sub-structure effect, the lower bound of $\Lambda \gtrsim 10^3$-10^4TeV can be obtained. However, this bound is also very much model-dependent. In fact, in some model calculations, the sub-structure effect turns out not to be of order m_ℓ/Λ but to be of order $(m_\ell/\Lambda)^2$, in which case the only much weaker lower bound of order 100GeV-1TeV can be obtained.

An even much larger lower bound of $\Lambda \gtrsim M_{GU} \sim 10^{14}$-$10^{16}$GeV, where M_{GU} is the mass scale of grand unification in grand unification gauge theories, may be obtained from a wishful thinking that all the features of grand unification gauge theories be preserved.

An expectation of the largest mass scale for the sub-structure comes from a (pregauge[17] and pregeometric[19]) supergrand unified subquark model[12] in which not only strong and electroweak interactions but also gravity appears as an effective interaction. Suppose that the fundamental length inverse (Λ which is presumably of order of the Planck mass) and the subquark mass (M) are both extremely large and of the same order. Then, it can be shown in a model[20] that the gauge coupling constant (e) and the Newtonian gravitational constant (G_N) are given by these large mass scales as

$$\frac{1}{4e^2} \sim \frac{\Lambda^4}{M^4} \quad \text{and} \quad \frac{1}{16\pi G_N} \sim \frac{\Lambda^4}{M^2}$$

so that they are related as

$$e^2 = 16\pi G_N M^2.$$

This relation suggests that the subquark mass may be of order $(\alpha/G_N)^{1/2}$, which is as large as 10^{18}GeV. Notice that a similar relation has been derived in Kaluza-Klein theories.[21]

From these discussions it has become clear that there seems no definite criterion yet to determine the mass scale for the substructure. I just hope that it is small enough (10^2GeV-10^2TeV) to be found in experiments by accelerators such as LEP and SSC which will be constructed in the near future.

5. Quarks and Leptons as Nambu-Goldstone Fermions

In order to explain the gross feature of the mass spectrum that the quark and lepton masses are much smaller than their size inverses, the following three possibilities have been proposed so far: quarks and leptons are 1) chiral fermions,[22] 2) Nambu-Goldstone fermions[7] or 3) "quasi-Nambu- Goldstone fermions"[23] although the second and third possibilities may not be independent. Very lately Mohapatra, Pati and Yasuè have given us some detail on the third possibility in Ref. 24. In this section, I shall discuss the second possibility that quarks and leptons are composite Nambu-Goldstone fermions of spontaneously broken supersymmetry in some detail in the minimal subquark model.[25]

Although the subquark dynamics is unknown, it may respect supersymmetry with the supercurrent[26]

$$s_\mu = M_w \gamma_\mu wC - i\gamma_\nu \gamma_\mu w\partial^\nu C.$$

If the supersymmetry is spontaneously broken, there will appear massless Nambu-Goldstone fermions. I shall take these massless Nambu-Goldstone fermions as the idealized quarks and leptons of the first generation (ν_e,e,u,d). The non-vanishing but small masses of e, u and d can be caused by the small breaking of supersymmetry. Such small breaking is parametrized by the non-vanishing $M_w - M_C$. Notice that for $M_w - M_C \neq 0$,

$$\partial^\mu s_\mu = -i(M_w^2 - M_C^2)wC.$$

It seems then natural to assume that the supersymmetric current is partially conserved as

$$\partial^\mu s_\mu \cong -iF_f^2 m_f f \quad \text{or} \quad s_\mu \cong F_f^2 \gamma_\mu f$$

where f is the quark or lepton field, m_f is the mass and F_f is the "decay constant". This partially conserved supercurrent (PCSC) hypothesis leads to the following formula for the quark or lepton mass

$$\begin{aligned} m_f &= F_f^{-4} \langle 0|\{\bar{S},[S,H(0)]\}|0\rangle \\ &= (M_w^2 - M_C^2)F_f^{-4}\langle \bar{w}w + M_w C^+ C\rangle_0 \quad \text{for} \quad H = M_w \bar{w}w + M_C^2 C^+ C \end{aligned}$$

where S is the supercharge. This result indicates that the quark and lepton masses may have the following properties: 1) They are proportional to the small parameter of supersymmetry breaking, $M_w^2 - M_C^2$. 2) For fixed $M_w^2 - M_C^2$, they may become smaller as Λ^{-1} as the mass scale of subquark dynamics or the size inverse of quarks and leptons (Λ) increases since it may be that $F_f \sim O(\Lambda)$ and $\langle \bar{w}w + M_w C^+ C\rangle_0 \sim O(\Lambda^3)$. This is on the contrary to a naive expectation. And 3) if the dynamical quantities F_f and $\langle \bar{w}w + M_w C^+ C\rangle_0$ are rather universal in a generation of quarks and leptons, the quark and lepton masses should satisfy the relation of

$$m_{\nu_e} - m_e \cong m_u - m_d \,.$$

This relation, however, is not well satisfied by the experimental values for the lepton masses ($m_{\nu_e} \cong 0$ and $m_e \cong 0.5$ MeV) and the estimates for the current quark masses ($m_u \cong$ 4-5MeV and $m_d \cong$ 8-9 MeV)[27] although the signs of both hand sides coincide. There are two ways out from this unsatisfactory feature of this relation. The first way is to discard the universality on which the relation is based and to proceed to difficult calculations of the dynamical quantities of F_f and $<\bar{w}w+M_w C^+C>_0$. The second way is to suppose that the subquark masses are all equal, i.e. $m_w = m_C$. Then, the quarks and leptons of the first generation are massless to the lowest order of supersymmetry breaking. Their masses may appear as higher order corrections of a supersymmetry breaking which lies in the subquark dynamics. Although the subquark interactions are totally unknown, their residual interactions at the quark and lepton level are known as ordinary strong and electroweak interactions which break supersymmetry. Therefore, the quark and lepton masses can be estimated by calculating the self-masses of quarks and leptons due to the residual interactions at the quark and lepton level. The electromagnetic self-mass of the electron, for example, is given approximately by

$$m_e^{(\gamma)} \cong \frac{-e^2}{F_e^4}\int\frac{d^4q}{(2\pi)^4}\,D^{\mu\nu}(q)\int d^4x\; e^{iqx}<0|T\{\bar{s}_\mu(x),s_\nu(0)\}|0>$$

$$\cong \frac{\alpha}{\pi}\int dm^2\rho_2^{(e)}(m^2)\,\ell n\,\frac{\Lambda_e^2+m^2}{m^2}$$

where $\rho_2^{(e)}$ is the electron spectral function and Λ_e is the size inverse of the eletron. Assume for simplicity that the $\rho_2^{(e)}$ is dominated by the electron and a possible excited electron, i.e., $\rho_2^{(e)}(m^2) = \kappa^2 m_e\delta(m^2-m_e^2)+\lambda^2 m_{e*}\delta(m^2-m_{e*}^2)$ where $\kappa^2 \cong 1$ and $\lambda^2 << 1$. Then, the only physically interesting case is that there exists an excited electron whose mass and coupling are large enough to satisfy

$$m_e \cong \frac{\alpha\lambda^2}{\pi}\,m_{e*}\,\ell n\,\frac{\Lambda_e^2+m_{e*}^2}{m_{e*}^2}$$

Such an excited electron may contribute to the anomalous magnetic moment of the electron. Suppose that the effective interaction between the electron and the excited electron is described by[13]

$$L_{e*} = \frac{e\lambda}{2m_{e*}}\,\bar{e}\sigma_{\mu\nu}e^*F^{\mu\nu} + \text{h.c.}$$

The contribution of the excited electron to the electron g-2 is estimated to be

$$\Delta a_{e*} \cong -\frac{9\alpha\lambda^2 m_e}{2\pi m_{e*}}\,\ell n\,\frac{\Lambda_e^2+m_{e*}^2}{m_{e*}^2}$$

By using the previous relation, this can be reduced to

$$\Delta a_{e*} \cong -\frac{9}{2}\left(\frac{m_e}{m_{e*}}\right)^2 .$$

By comparing this with the most precise comparison between the experimental and theoretical values for the electron g-2,[16] the mass of the excited electron can be estimated to be $m_{e*} \cong (81 \pm 27)$ GeV. If the logarithmic factor $\ln(\Lambda_e^2 + m_{e*}^2)/m_{e*}^2$ is of order unity, which seems reasonable, the coupling constant can also be estimated to be $\lambda^2 \cong (1.8 \pm 0.6) \times 10^{-3}$. This picture of the composite electron as a Nambu-Goldstone fermion suggests that there may exist an excited electron whose mass lies between 50GeV and 110GeV and whose coupling constant (λ^2) lies between 10^{-3} and 10^{-2}. The possible existence of such an excited electron is not only consistent with the existing e^+e^- colliding beam experimental data[8] but suitable for explaining the anomalous events lately reported by the UA1 and UA2 experiments at SPS $\bar{p}p$ Collider for production of Z bosons[28] although the strongly interacting, composite or excited weak boson is another promising explanation.[29] Notice that the decay width of the excited electron is large enough, i.e. $\Gamma(e^* \to e\gamma) \cong \alpha\lambda^2 m_{e*}/2 \cong (0.15 \pm 0.09)$MeV although the coupling constant λ is small.

In the above considerations, possible contributions to the electron mass from the weak interactions are ignored since they are small. It is, however, clear that the quark mass receives a dominant contribution from the gluon since the gluon coupling constant is much larger than the electromagnetic one. Therefore, it is natural to expect that the ratio of the up or down quark mass to the electron mass becomes of order α_s/α if excited quarks have their masses and couplings similar to the excited eletron's. This expectation seems satisfactory since it provides an explanation of the fact that the up and down quark masses are roughly by ten times larger than the eletron mass. A similar conclusion has recently been reached by Yasuè from different consideration in a different composite model of quarks and leptons.[30]

6. 't Hooft Condition

Since 't Hooft proposed the condition that the sum of the Adler-Bell-Jackiw axial-vector triangle anomalies due to light composite fermions be matched with the sum of those due to constituent fermions,[22] many people have taken it seriously to restrict the possible type of composite models or to determine the number of generations in the models. In this section, I shall point out that the anomaly matching condition should not be taken seriously if not only quarks and leptons but also gauge bosons are composite as in our grand unified subquark models. The reason for this is simple and the following: The condition has been derived from the assumed renormalizability of both the gauge theory at the quark and lepton level and the gauge theory at the subquark level, which requires the anomaly free conditions at both levels. The gauge fields have been assumed to be elementary at both levels. Furthermore, the presence of "spectator fermions" which contribute to the anomaly at both levels equally has been assumed. Therefore, even if the presence of such spectator fermions (which is doubtful) is accepted, the anomaly matching condition would not be borne out since the gauge fields which look elementary at the quark and lepton level disappear at the subquark level if they are composite. The assumed renormalizability only requires the anomaly free condition at the quark and lepton level for the composite gauge bosons and the anomaly free condition at the subquark level for the elementary gauge bosons (subgluons), if any. If one wishes to preserve the anomaly matching condition even in composite models where gauge bosons are also composite, one must appeal to some other principle than renormalizability. An example for such principle is "duality" that some physical quantities can be evaluated equally both at the composite level and at the constituent level.

7. "Duality"

Such possible "duality" can be claimed for the composite gauge bosons. For example, the weak bosons and the Higgs scalars can be taken not only as a composite of wakem-antiwakem pair but also as a composite of lepton-antilepton or quark-antiquark pair. If this is the case, the previously discussed sum rules[2,12] for the subquark masses and for the lepton and quark masses may simultaneously hold as

$$\sqrt{\Sigma m_w^4/\Sigma m_w^2} = \sqrt{\Sigma m_f^4/\Sigma m_f^2} = m_H/2$$

$$\gtrsim \sqrt{\langle m_w^2\rangle} = \sqrt{\langle m_f^2\rangle} = m_{W^\pm}/\sqrt{3} \cong 46\text{GeV}.$$

8. Flavor Mixing

There are three ways to understand why the Cabibbo-GIM-KM quark mixing[31] occurs for the weak charged current. The first one is the "hakam mixing", which means that quark mixing is caused by the intrinsic mixing of "hakams" h_i $(i=1,2,\cdots,N)$, the subquarks which have horizontal gauge quantum numbers or generation numbers. If this is the case, since a hakam is shared by quarks and leptons of the same generation, the mixing angles for quarks and leptons would become of the same order of magnitude. This possibility would become relevant if such large lepton mixing is found in neutrino oscillation experiments. However, no definite evidence for neutrino oscillation has yet been reported.[32]

The second possibility is "level mixing",[33] which seems more natural. The weak charged current which has been written in terms of hadrons and in terms of quarks can be most fundamentally written in terms of subquarks as[12]

$$J_\mu = \frac{G^\beta}{G^\mu}\bar{p}\gamma_\mu(1 - \frac{g_A^\beta}{g_V^\beta}\gamma_5)n + \frac{G^\Lambda}{G^\mu}\bar{p}\gamma_\mu(1 - \frac{g_A^\Lambda}{g_V^\Lambda}\gamma_5)\Lambda + \cdots$$

$$= U_{ud}\bar{u}\gamma_\mu(1-\gamma_5)d + U_{us}\bar{u}\gamma_\mu(1-\gamma_5)s + \cdots$$

$$= \bar{w}_1\gamma_\mu(1-\gamma_5)w_2 \ .$$

Let us also suppose that the origin of quark (and lepton) generation is dynamical. In other words, a quark (or lepton) of a higher generation (u_n or d_n for $n>1$) is considered to be an excited state of its corresponding one of the lowest generation (u or d). Then, the mixing matrix of quarks U_{mn} can be defined by the matrix element of the subquark current between the m-th up-like quark and the n-th down-like quark as[33]

$$\langle u_m|\bar{w}_1\gamma_\mu w_2|d_n\rangle = U_{mn}\bar{u}_m\gamma_\mu d_n \ .$$

An immediate consequence of this picture is that the mixing matrix elements may vary as functions of momentum transfer between quarks. The algebra of subquark current includes the familiar commutation relation of isospin current,[7]

$$\delta(x_0-y_0)[V_0^+(x),V_0^-(y)] = 2\delta^4(x-y)V_0^3(x) \ .$$

This relation sandwiched between quark states leads to the unitary of the quark mixing matrix, i.e.[34]

$$UU^\dagger = U^\dagger U = 1$$

if the intermediate states form a complete set as

$$|u_\ell\rangle\langle u_\ell| = 1 \qquad \text{and} \qquad |d_\ell\rangle\langle d_\ell| = 1 \ .$$

Furthermore, if the isospin breaking is perturbative, the mixing matrix elements are given by

$$U_{mn} = \frac{<u_m|H_I|u_n>}{m_{u_m}-m_{u_n}} + \frac{<d_m|H_I|d_n>}{m_{d_n}-m_{d_m}} \quad \text{for} \quad m \neq n.$$

This indicates that the off-diagonal mixing matrix elements between different generations decrease as fast as or faster than the inverse of mass difference between the relevant quarks. It also indicates that the off-diagonal mixing matrix elements have the antisymmetric property of

$$U_{mn} = -U^*_{nm} \ .$$

This property is in excellent agreement with the recent analysis of the experimental data made by Chau and also by Lee-Franzini,[35] which concludes $U^{exp}_{us} \cong 0.23$ and $U^{exp}_{cd} \cong -0.24$. These properties of the quark mixing matrix as well as the above mentioned automatic unitarity of it seem to be very natural and can be taken as a successful consequence of the subquark model.

Furthermore, the quark mixing matrix and its momentum dependence have been explicitly calculated by Akama and myself and, independently, by Tomozawa in naive potential models for the subquark-binding force.[36] The results for the quark mixing matrix in case of the square-well potential, for example, shown by

$$U_{mn} \cong \sqrt{\rho}\,\frac{2m\ \sin(n\rho-m)\pi}{(n^2\rho^2-m^2)\pi}$$

$$= \begin{bmatrix} 0.9737\pm0.0025(\text{input}) & 0.20\pm0.01 & -0.092\pm0.003 & \cdots \\ -0.14\pm0.01 & 0.903\pm0.009 & 0.37\pm0.01 & \cdots \\ 0.081\pm0.003 & -0.21\pm0.01 & 0.79\pm0.02 & \cdots \end{bmatrix},$$

where ρ is a single parameter indicating the order of isospin breaking ($\rho = 0.878\pm0.006$), were roughly consistent with the experimental data.[37] However, as discussed in detail by Chau, and Lee-Franzini,[35] the latest data for the lifetime of B-meson show $|U^{exp}_{cb}| = 0.0435\pm0.0047$, which strongly disagrees with the above results. A possible way out is to discard the naive potential model and to go back to the perturbative picture where

$$\frac{|U_{cb}|}{|U_{us}|}\left(\cong \frac{|U_{ts}|}{|U_{cd}|}\right) \cong \frac{m_s}{m_b}\cdot\frac{<s|H_I|b>}{<d|H_I|s>} \qquad \text{for} \quad m_s \ll m_c \ll m_b \ll m_t$$

$$\cong \frac{m_s}{m_b} \quad \text{if } |\langle s|H_I|b\rangle / \langle d|H_I|s\rangle| \cong 1 .$$

This relation indicates that U_{cb} can be enough (possibly too much) suppressed by the factor of m_s/m_b (= 0.03∿0.1) compared to U_{us}. Furthermore, if the matrix elements of the perturbative Hamiltonian (H_I) between quark states whose generation difference is larger than one vanish, which may likely happen due to some quantum number conservation, the quark mixing matrix elements U_{ub} and U_{td} can appear as the second order perturbative effects and can be related to the other elements as

$$|U_{ub}| \cong \frac{m_s}{m_c}|U_{us}|\cdot|U_{cb}| \quad \text{if} \quad |\langle u|H_I|c\rangle / \langle d|H_I|s\rangle| \cong 1$$

and

$$|U_{td}| \cong |U_{us}|\cdot|U_{cb}|$$

The first relation indicating $|U_{ub}|/|U_{cb}| \cong 0.03\sim0.08$ is perfectly consistent with the experimental upper bound of $|U_{ub}|/|U_{cb}| \leq 0.119$.[35,37] The second relation predicts $|U_{td}| \cong 0.01$, which can be checked by future experimental result.

Although we have lost some reliability on the naive potential model for the quark mixing matrix, it is still instructive to present the results for the momentum dependence of the individual matrix elements which seems less model-dependent. The results in case of the square-well potential, for example, are illustrated in Fig. 2, where x is the product of the momentum transfer and the

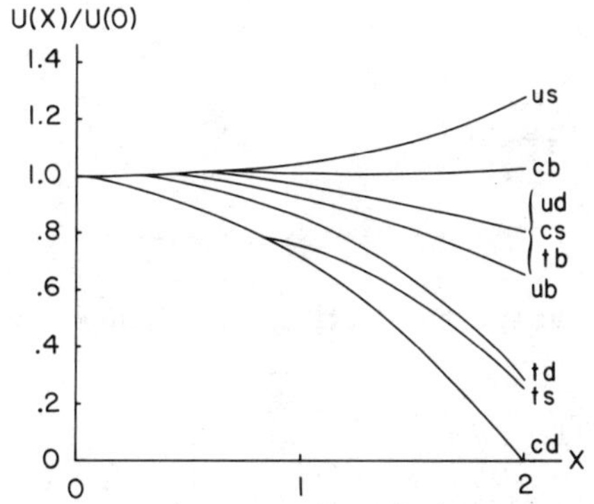

FIG.2. The momentum dependence of the quark mixing matrix elements.

quark size. These results show that the quark mixing matrix elements drastically change when the momentum transfer between quarks grows up to an order of the size inverse of quarks. We have suggested that this phenomenon of varying quark mixing matrix elements may be found in high energy neutrino reactions by measuring the ratio of strangeness changing to non-changing events or that of di-muon to single-muon events, etc.. Suppose that the semi-leptonic decays of topped mesons will be analyzed in detail in future e^+e^- colliding-beam experiments. It will then be possible to measure the mixing matrix elements $U_{ti}(x)$ (i=d,s,b) for fairly large momentum transfers between quarks and to find the momentum-transfer dependence of them. In any case, such possible discovery of the momentum dependence of the quark mixing matrix elements would provide one of the most eminent signs for the sub-structure of quarks.

The third possibility is "combinations".[38] If in a composite model there exist more than one composite states which have the same quantum numbers as a quark or lepton, due to different combinations of subquarks, they can be assigned to quark or lepton states of different generations. For example, as mentioned earlier, in the minimal composite model there exist three (or four) generations of composite quarks and leptons corresponding to (wC), $(\overline{www}C)$ and $(wC^+C^+C^+)$ states. The quark mixing in this picture of generations is caused by the condensation of (wwww) or (CCCC) (See Fig. 3). If the mixings due to $\langle wwww\rangle_0$ and $\langle CCCC\rangle_0$ are of order ε

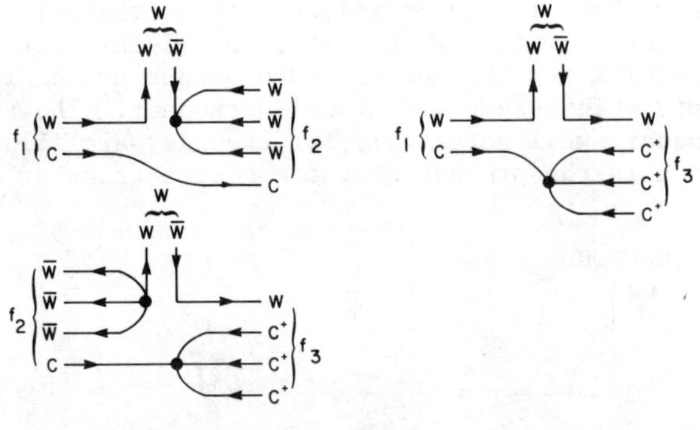

FIG.3. The subquark diagrams for quark mixing.

and η $(\varepsilon,\eta \ll 1)$ respectively, the mixing matrix has the structure of either one of

$$|U_{mn}| = \begin{pmatrix} 1 & \varepsilon & \eta \\ \varepsilon & 1 & \varepsilon\eta \\ \eta & \varepsilon\eta & 1 \end{pmatrix}, \begin{pmatrix} 1 & \varepsilon & \varepsilon\eta \\ \varepsilon & 1 & \eta \\ \varepsilon\eta & \eta & 1 \end{pmatrix}, \begin{pmatrix} 1 & \varepsilon\eta & \varepsilon \\ \varepsilon\eta & 1 & \eta \\ \varepsilon & \eta & 1 \end{pmatrix}, \quad \text{or} \quad (\varepsilon \leftrightarrow \eta) ,$$

depending on which one of (wC), $(\overline{www}C)$ and $(wC^+C^+C^+)$ corresponds to the first, second and third generation of quarks and leptons.

Since the experimental data[35,37] indicates that $|U_{cb}|^2 \gg |U_{ub}|^2$, only the second or fifth case where

$$|U_{ub}| = |U_{us}|\cdot|U_{cb}| \cong 0.01$$

and

$$|U_{td}| = |U_{ts}|\cdot|U_{cd}|$$

can survive. However, this is a rather uncomfortable case where the simplest (wC) states correspond not to the first generation but to the second. There have been proposed more sofisticated models which produce more satisfactory structures of the quark mixing matrix.[38]

9. Excited Quarks, Leptons and Gauge Bosons

Recently, it has been emphasized that the SPS $\bar{p}p$ Collider data from the UA1 and UA2 groups for Z production contain the anomalous events:[28] one of the several UA1 "e^+e^-" events, one of the few UA1 "$\mu^+\mu^-$" events and one of the several UA2 "e^+e^-" events are actually an $e^+e^-\gamma$ event, a $\mu^+\mu^-\gamma$ event and an e^+e^- event, respectively, in which a very hard γ (the energy $\gtrsim$ 10GeV) is associated with a lepton-antilepton pair. Very lately Yamada has given us a review on the experimental data in more detail in Ref. 11. These anomalous events can not be explained by either one of 1) (internal or external) hard bremsstrahlung, 2) Higgs production followed by its decay into 2γ, 3) toponium production followed by its decay into $\ell^+\ell^-$ or 4) heavy lepton production followed by its decay into $\ell\pi^0$ (See Fig. 4). The much more promising explanations of these are either 5) excited-lepton production followed by its decay into $\ell\gamma$ or 6) $\gamma Z'$ production followed by Z' decay into $\ell^+\ell^-$ (the invariant mass of order 50GeV) where Z' is another (real or virtual) (pseudo scalar or excited vector) state of Z (See Fig. 4).

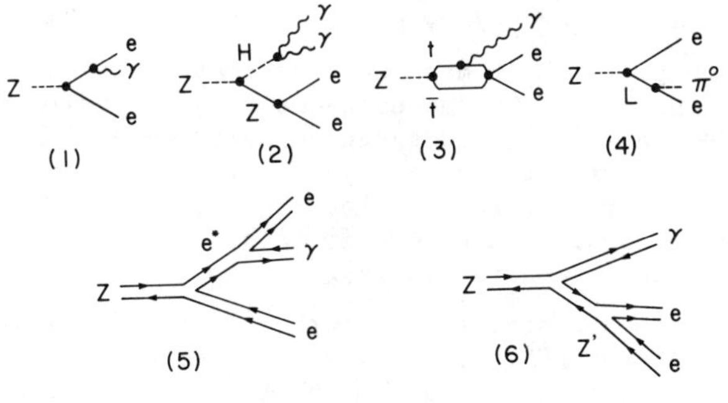

FIG.4. Various processes for the anomalous "e^+e^-" events.

If the anomalous events are due to the cascade decay of Z into an excited lepton and a lepton followed by the decay of the excited lepton into a lepton and a photon, the mass of the excited lepton is estimated to be either about 9GeV or about 75GeV, depending on which lepton is associated with the photon. As emphasized in Section 5, the existence of such light excited leptons is yet consistent with the other existing data[8] if the photon coupling is sufficiently small. It even agrees with the estimation of $m_{e*} \cong (81\pm27)$GeV in the picture of quarks and leptons as composite Nambu-Goldstone fermions discussed in Section 5.

As has been emphasized many times,[39] the phenomenological success of Glashow-Salam-Weinberg theory of electroweak interactions for describing low energy phenomena does not necessarily mean that the gauge bosons $W^{\pm}$ and Z are elementary. There may be many other states and many excited gauge bosons in the $W^{\pm}$ and Z channels. All that are necessary for preserving the success of the G-S-W theory are the sum rules, for example, for "Z" production by e^+e^- colliding beams

$$\int ds \, \frac{\sigma(e^+e^- \text{"Z"})}{s} = \frac{\pi G_F}{2\sqrt{2}} \, [1+(4\sin^2\theta_W)^2]$$

and

$$\int ds \, \sigma(e^+e^- \to \text{"Z"}) = \frac{\pi^2\alpha}{4\sin^2\theta_W \cos^2\theta_W} \, [1+(1-4\sin^2\theta_W)^2] \; .$$

Therefore, it is perfectly possible that the existence of relatively light composite bosonic states and excited gauge bosons explains not only these anomalous $e^+e^-\gamma$ (or $\mu^+\mu^-\gamma$) events but also the anomalous $W^{\pm}$ jet events with the invariant masses of "heavy objects" of order 160-186GeV and the anomalous two-jet events with the invariant masses of "heavy objects" of order 147 ± 3GeV reported by the UA2 group.[28] Their data suggest that there may exist a composite bosonic state of subquark-antisubquark pair with the mass of order 50GeV and an excited gauge bosonic state of subquark-antisubquark pair with the mass of order 140-190GeV, which remarkably fits our expectation (See Fig. 5).

In either case of 5) and 6), it is highly desirable to observe direct production of excited quarks, leptons, excited gauge bosons and composite bosonic states in future e^+e^- colliding beam experiments at TRISTAN and LEP. For a moment, however, it is worth watching whether the rather large (about 1/4-1/8) ratio of the anomalous $e^+e^-\gamma$ (or $\mu^+\mu^-\gamma$) events for Z production and the rather

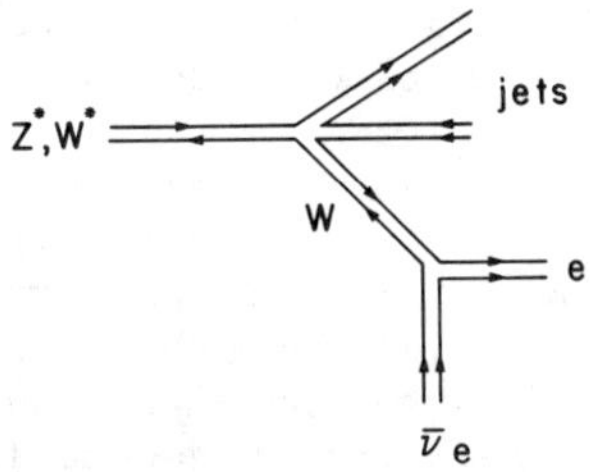

FIG.5. The subquark diagram for the anomalous $W^{\pm}$ jet events.

large rate of the anomalous $W^{\pm}$ jet and two-jet events for "heavy object" production will remain with higher statistics. If they remain, it may indicate a great discovery of the sub-structure which is even more exciting than that of $W^{\pm}$ and Z.

10. Half-Charged Particles

Whether subquarks are permanently confined in a quark or a lepton depends on the unknown subquark dynamics. If they are not confined, their fractional charges can be an excellent sign of their presence. In the minimal composite model, the charges of w_1, w_2 and C_0 are +1/2, -1/2 and -1/2, respectively. The charge of the color singlet state $(C_1C_2C_3)$ is also +1/2. Therefore, a possible discovery of the half-charged particles would be a dramatic sign of the subquarks. Such a search for half-charged particles is under way. In the underground search by Orito et al.,[40] no anomalous candidate has yet been found and the upper bound on the flux of $1.6\times10^{-12}\mathrm{cm}^{-2}\mathrm{s}^{-1}\mathrm{sr}^{-1}$ has been reported. On the other hand, I have heard that the group of Fairbank who claimed the discovery of the fractionary charges of $\pm2/3$ or $\pm1/3$[41] has found the half charge also. It is highly desirable to receive more detailed information.

11. Cosmology with Composite Quarks and Lepton

Finally, let me introduce one of the most interesting contributed papers to the XXII International Conference on High Energy Physics. It is written by Chikashige and Ito on cosmology in composite models.[42]

Apart from the terrestrial experiments discussed in the previous sections, the early stage of our universe serves as a good laboratory for the study of the sub-structure of quarks and leptons. When the temperature T is of order Λ, the distance between quarks and leptons was comparable to the size of these particles. Therefore, at the era with $T\sim\Lambda$, the finiteness of the size of quarks and leptons was so significant that it must have affected the evolution of the universe. To see this, they have assumed the van der Waals type of equation of state for the quark-lepton gas at this era.

As is well known, in the van der Waals equation of state, $\partial p/\partial V$ may become positive below a critical temperature, which leads to a first-order phase transition. If the barrier for the nucleation of bubbles was so high that the transition was delayed, the enormous entropy due to reheating might have been produced at this era. This is a new version of the scenario for the inflationary universe[43] in composite models.

Another interesting possibility in cosmology in composite models is to explain the baryon number asymmetry in the present universe. There are two solutions for this problem: the baryon-antibaryon symmetric cosmological model by Omnès[44] and the asymmetric one by Sakharov.[45]

In the symmetric cosmological model, the baryon number asymmetry is explained by assuming the separation of matter and antimatter at the early stage of the universe. It was suggested in QCD that such a separation might have occurred at the quark-lepton era. If the subquark dynamics can be described by QSCD, it must also have caused the separation of subquarks and anti-subquarks at the subquark era. In fact, the QCD force would not be enough to explain the baryon number asymmetry in the symmetric cosmological model. Suppose the subquark dynamics does not contribute to the separation of matter and antimatter. Then, only C and $\bar{C}$ were separated due to the QCD force, but $\bar{w}$ and w must have been mixed together. After the universe was cooled down to to the temperature below $T \sim \Lambda$, C and w were bound to form quarks and leptons. In this case, however, the quark number density must have been rather small since it was determined by the number density of w subquark before the formation of quarks and leptons. Therefore, the symmetric cosmological model will survive if the subquark force is described by such a dynamics as QSCD, which can separate subquarks and anti-subquarks.

As I discussed in the previous section, the baryon number violating process can occur in subquark models. The baryon number asymmetry in the present universe may be explained if baryon number conservation and CP are violated simultaneously in the decay of heavy composite bosons such as the X boson.

Also, the condensation of C subquarks can contribute to the baryon number violation. However, in this case there seems to be a long way before calculating the surviving baryon number asymmetry in the present universe. How to calculate it seems to require further investigations.

12. Conclusion

In concluding this chapter, I would like to emphasize that the idea of composite models of quarks and leptons (and also gauge bosons as well as Higgs scalars) which was proposed by us, theorists, in the middle of seventies have just become a subject of experimental relevance in the middle of eighties.

Note added:

Some contents of this chapter overlap those of the previous chapter. However, to repeat is the best exercise, in any case.

References and Footnotes

1. For a review of some "prehistorical" works, see, for example, H. Terazawa, Phys. Rev. D22 (1980) 184. For some earlier reviews, see, for example, H. Terazawa, in Proc. XIX Int. Conf. on High Energy Physics, Tokyo, Aug. 23-30, 1978, edited by S. Homma et al. (Physical Society of Japan, Tokyo, 1979), p.617, Proc. Japan-Italy Symp. on Fundamental Physics, Tokyo, Jan. 27-30, edited by S. Fukui and T. Toyoda (Nagoya Univ., Nagoya, 1981), p.65 and in Proc. 1981 INS Symp. on Quark and Lepton Physics, Tokyo, June 25-27, 1981, edited by K. Fujikawa et al. (INS, Univ. of Tokyo, Tokyo, 1981), p.296, M.E. Peskin, in Proc. 1981 Int. Sym. on Lepton and Photon Interactions at High Energies, Bonn, Aug. 24-19, edited by W. Pfeil (Univ. Bonn, Bonn, 1981), p.880, H. Terazawa, in Proc. XXI Int. Conf. on High Energy Physics, Paris, July 26-31, 1982, edited by P. Petiau and M. Porneuf, J. de Phys. C3, 289 (1982), L. Lyons, Prog. Part. Nucl. Phys. 10, 227 (1983), R. Barbieri, in Proc. 1983 Int. Symp. on Lepton and Photon Interactions at High Energies, Cornell, Aug. 4-9, 1983, edited by D.G. Cassel and D.L. Kreinick (LNS, Cornell Univ., Ithaca, N.Y., 1983), p.479. For some latest reviews, see, for example, H. Terazawa, in Proc. Europhysics Conf. on Flavor Mixing in Weak Interactions, Erice, March 4-12, 1984, edited by L.-L. Chau (Plenum Pub. Co., New York, 1984), p.655 and in Proc. XXII Int. Conf. on High Energy Physics, Leipzig, July 19-25, 1984, edited by A. Meyer and E. Wieczorek (Akademie der Wissenschaften der DDR, Zeuthen, 1984), Vol.I, p.63. Naturally, many parts of the content of this review have already appeared in the last literature.
2. H. Terazawa, Y. Chikashige and K. Akama, Phys. Rev. D15, 480 (1977).
3. J.C. Pati and A. Salam, Phys. Rev. D10, 275 (1974).
4. H. Georgi and S.L. Glashow, Phys. Rev. Lett. 32, 438 (1974). The possible proton decay had been disscussed earlier by Yamaguchi and by Pati and Salam. See Y. Yamaguchi, Prog. Theor. Phys. Suppl. 11, 33 (1959) and J.C. Pati and A. Salam, in Ref. 3.
5. For the latest review, see M. Koshiba, in Poc. XXII Int. Conf. on High Energy Physics, Leipzig, July 19-25, 1984, edited by A. Meyer and E. Wieczorek (Akademie der Wissenschaften der DDR, Zeuthen, 1984), Vol.II, p.67.
6. H. Harari, Phys. Lett. 86B, 83 (1979); M.A. Shupe, Phys. Lett. 86B, 87 (1979).

7. H. Terazawa, Prog. Theor. Phys. 64, 1763 (1980).
8. For some of the latest reviews, S. Yamada, in Proc. 1983 Int. Symp. on Lepton and Photon Interactions at High Energies, Cornell, Aug. 4-9, 1983, edited by D.G. Cassel and D.L. Kreinick (LNS, Cornell Univ., Ithaca, N.Y., 1983), p.525 and A. Silverman, in Proc. XXII Int. Conf. on High Energy Physics, Leipzig, July 19-25, 1984, edited by A. Meyer and E. Wieczorek (Akademie der Wissenschaften der DDR, Zeuthen, 1984), Vol.II, p.91.
9. H. Terazawa, Phys. Rev. D22, 2921 (1980).
10. For the latest reviews, see R.K. Böck and A. Silverman, in Proc. XXII Int. Conf. on High Energy Physics, Leipzig, July 19-25, 1984, edited by A. Meyer and E. Wieczorek (Akademie der Wissenschaften der DDR, Zeuthen, 1984), Vol.II, p.2 and p.91.
11. S. Yamada, in Proc. XXII Int. Conf. on High Energy Physics, Leipzig, July 19-25, 1984, edited by A. Meyer and E. Wieczorek (Akademie der Wissenschaften der DDR, Zeuthen, 1984), Vol.I, p.72.
12. H. Terazawa, Phys. Rev. D22, 184 (1980).
13. F.E. Low, Phys. Rev. Lett. 14, 238 (1965); H. Terazawa, Prog. Theor. Phys. 37, 204 (1967); H. Terazawa, M. Yasuè, K. Akama and M. Hayashi, Phys. Lett. 112B, 387 (1982).
14. E.J. Eichten, K.D. Lane and M.E. Peskin, Phys. Rev. Lett. 50, 811 (1983).
15. For example, see H. Terazawa et al., in Ref. 13 and references therein.
16. T. Kinoshita and W.B. Linquist, Phys. Rev. Lett. 47, 1573 (1981); T. Kinoshita, B. Nižic and Y. Okamoto, CLNS-83/592 (Cornell Univ.) Dec., 1983.
17. J.D. Bjorken, Ann. Phys. 24, 174 (1963); H. Terazawa et al., in Ref. 2. See also H.B. Nielsen, in Ref. 18.
18. H.B. Nielsen, "Field Theories without Fundamental Gauge Symmetries", a preprint (The Niels Bohr Inst.) 1984.
19. A.D. Sakharov, Doklady Akad. Nauk SSSR 177, 70 (1969) [Sov. Phys. JETP 12, 1040 (1969)];
K. Akama, Y. Chikashige, T. Matsuki and H. Terazawa, Prog. Theor. Phys. 60, 868 (1978) and some earlier references therein.
20. H. Terazawa, Phys. Lett. 133B, 57 (1983). For some related works, see H. Terazawa and K. Akama, Phys. Lett. 96B, 276 (1980), 97B, 81 (1980) and H. Terazawa, Phys. Lett. 101B, 43 (1981).
21. T. Kaluza, Sitzungsber. Preuss. Akad. Wiss. Kl, 966 (1921); O. Klein, Z. Phys. 37, 895 (1926).
22. G. 't Hooft, in Recent Developments in Gauge Theories, edited by G. 't Hooft et al. (Plenum Pub. Co., New York, 1980), p.135.
23. W. Buchmüller, R.D. Peccei and T. Yanagida, Phys. Lett. 124B, 67 (1983); R. Barbieri, A. Masiero and G. Veneziano, Phys. Lett. 128B, 179 (1983); O.W. Greenberg, R.N. Mohapatra and M. Yasuè, Phys. Lett. 128B, 65 (1983).

24. R.N. Mohapatra, J.C. Pati and M. Yasuè, Phys. Lett. 151B, 251 (1985).
25. For more details, see H. Terazawa, INS-Report-485 (INS, Univ. of Tokyo) Dec., 1983. A similar discussion has been presented in W. Bardeen and V. Višnjić, Nucl. Phys. B194, 422 (1982) and W.A. Bardeen, T.R. Taylor and C.K. Zachos, Nucl. Phys. B231, 235 (1984).
26. H. Miyazawa, Prog. Theor. Phys. 36, 1266 (1966); J. Wess and B. Zumino, Nucl. Phys. B70, 39 (1974).
27. For example, see S. Weinberg, Transactions of the New York Academy of Sciences, Vol.38 (1977).
28. G. Arnison et al. (UA1), Phys. Lett. 126B, 398 (1983), P. Bagnaia et al. (UA2), Phys. Lett. 129B, 130 (1983) and Phys. Today Vol. 36, No.11, 17 (1983). See also P. Hansen (UA2), in Proc. Europhysics Conference on Flavor Mixing in Weak Interactions, Erice, March 4-12, 1984, edited by L.-L. Chau (Plenum Pub. Co., New York, 1984), J. Rohlf (UA1) and J.-P. Repellin (UA2), in Proc. XXII Int. Conf. on High Energy Physics, Leipzig, July 19-25, 1984, edited by A. Meyer and E. Wieczorek (Akademie der Wissenschaften der DDR, Zeuthen, 1984), Vol.II, p.12 and p.21
29. H. Terazawa, Phys. Rev. D7, 3663 (1973) and in Ref. 12.
30. M. Yasuè, Physics Publication No.66 (Univ. of Maryland) Oct., 1983, to be published in Nuovo Cim. A.
31. N. Cabibbo, Phys. Rev. Lett. 10, 531 (1963), S.L. Glashow, J. Iliopoulos and L. Maiani, Phys. Rev. D2, 1285 (1970), M. Kobayashi and T. Maskawa, Prog. Theor. Phys. 49, 652 (1973).
32. For some of the latest reviews, see M.H. Shaevitz, in Proc. 1983 Int. Symp. on Lepton and Photon Interactions at High Energies, Cornell, Aug. 4-9, 1983, edited by D.G. Cassel and D.L. Kreinick (LNS, Cornell Univ., Ithaca, N.Y., 1983), p.132 and B.A. Ljubimov, in Proc. XXII Int. Conf. on High Energy Physics, Leipzig, July 19-25, 1984 edited by A. Meyer and E. Wieczorek (Akademie der Wissenschaften der DDR, Zeuthen, 1984), Vol.II, p.108.
33. H. Terazawa, Prog. Theor. Phys. 58, 1276 (1977).
34. V. Višnjić-Triantafillow, FERMILAB-Pub.-80/34-THY (Fermilab, March, 1980), H. Terazawa, in Ref. 7, O.W. Greenberg and J. Sucher, Phys. Lett. 99B, 339 (1981).
35. For a review, see L.-L. Chau, Phys. Rep.Vol.95, No.1, 1 (1983). For some of the latest analyses, see L.-L. Chau, BNL-33951 (BNL) 1983 and J. Lee-Franzini, in Proc. Europhysics Conf. on Flavor Mixings in Weak Interactions, Erice, March 4-12, 1984, edited by L.-L. Chau (Plenum Pub. Co., New York, 1984) and in Proc. XXII Int. Conf. on High Energy Physics, Leipzig, July 19-25, 1984, edited by A. Meyer and E. Wieczorek (Akademie der Wissenschaften der DDR, Zeuthen, 1984), Vol.I, p.150.
36. H. Terazawa and K. Akama, Phys. Lett. 101B, 190 (1981); Y. Tomozawa, UM HE 81-3 (Univ. of Michigan) 1981, in Proc. 1981 INS Symp. on Quark and Lepton Physics, Tokyo, June 25-27,

1981, edited by K. Fujikawa et al. (INS, Univ. of Tokyo, Tokyo, 1981), p.319 and Phys. Lett. 104B, 136 (1981).

37. For some of the latest reviews, see S. Stone and N.W. Reay in Proc. 1983 Int. Symp. on Lepton and Photon Interactions at High Energies, Cornell, Aug. 4-9, 1983, edited by D.G. Cassel and D.L. Kreinick (LNS, Cornell Univ., Ithaca, N.Y., 1983), p.203 and p.244, G.H. Trilling and W.T. Ford, in Proc. Europhysics Conf. on Flavor Mixings in Weak Interactions, Erice, March 4-12, 1984, edited by L.-L. Chau (Plenum Pub. Co., New York, 1984), and P. Langacker, in Proc. XXII Int. Conf. on High Energy Physics, Leipzig, July 19-25, 1984, edited by A. Meyer and E. Wieczorek (Akademie der Wissenschaftern der DDR, Zeuthen, 1984), Vol.II, p.215. Also, for the new upper bound on the mass of quoted in Section 3, see G.H. Trilling, in this Ref. and B.A. Ljubimov, in Proc. XXII Int. Conf. on High Energy Physics, Leipzig, July 19-25, 1984, edited by A. Meyer and E. Wieczorek (Akademie der Wissenschaften der DDR, Zeuthen, 1984), Vol.II, p.108.

38. R. Casalbuoni, G. Domokos and S. Karesi-Domokos, Phys. Rev. D23, 462 (1981); K. Matumoto and K. Kakazu, Prog. Theor. Phys. 65, 390 (1981); M. Yasuè, Prog. Theor. Phys. 65, 1995 (1981); S. Weinberg, in Proc. Workshop on Weak Interactions as Probes of Unification, VPI, Dec. 4-6, 1980, edited by G.B. Collins et al. (AIP, New York, 1981), p.521.

39. H. Terazawa, in Refs. 1, 12, 29 and in Proc. XXI Int. Conf. on High Energy Physics, Paris, July 26-31, 1982, edited by P. Petiau and M. Porneuf, J. de Phys. C3, 289 (1982).

40. T. Mashimo, S. Orito, K. Kawagoe, S. Nakamura and M. Nozaki, Phys. Lett. 128B, 327 (1983).

41. G.S. Larue, J.D. Phillips and W.M. Fairbank, Phys. Rev. Lett. 46, 967 (1981).

42. Y. Chikashige and I. Ito, a contributed paper to the XII Int. Conf. on High Energy Physics, Leipzig, July 19-25, 1984.

43. For the latest review, see A.D. Linde, in Proc. XXII Int. Conf. on High Energy Physics, Leipzig, July 19-25, 1984, edited by A. Meyer and E. Wieczorek (Akademie der Wissenschaften der DDR, Zeuthen, 1984), Vol.II, p.125.

44. R. Omnès, Phys. Rev. Lett. 23, 38 (1969) and Phys. Rep. 3C, 1 (1972).

45. A.D. Sakharov, ZhETF Pis'ma 5, 32 (1967) [Sov. Phys. JETP Lett. 5, 24 (1967)] and ZhETF 76, 1172 (1979) [Sov. Phys. JETP 49, 594 (1979)], V.A. Kuzmin, ZhETF Pis'ma 13, 335 (1970), A. Yu Ignatiev, N.V. Krasnikov, V.A. Kuzmin and A.N. Tavkhelidze, Phys. Lett. 76B, 436 (1978), M. Yoshimura, Phys. Rev. Lett. 41, 281 (1978), 42, 746(E) (1979). A similar idea to explain the baryon number asymmetry in our universe had been presented earlier by Yamaguchi. See Y. Yamaguchi, in Ref. 4.

III. SUPERGRAND UNIFIED COMPOSITE MODEL

1. Introduction

It is a final goal in theoretical physics to construct the fundamental theory which can not only describe but explain all physical phenomena. Historically, many attempts have been made to construct such a final theory in physics. Notable among them are the unified field theory of Einstein, Weyl, Kaluza and Klein,[1] the unified spinor theory of Heisenberg, Ivanenko and Nambu-Jona-Lasinio,[2] the unified gauge theory of Schwinger, Glashow, Salam and Weinberg[3] and the grand unified gauge theory of Pati-Salam and Georgi-Glashow.[4] The first two of these have become obsolete in their original forms for obvious reasons while the last two are incomplete as the final theory since they can neither describe gravity nor explain all fundamental physical quantities such as the gauge coupling constants and the masses of the fundamental fermions (leptons and quarks) and the gauge bosons ($W^{\pm}$ and Z).

Recently, we[5] have proposed the supergrand unified pregauge[6] and pregeometric[7] subquark model[8] in which not only the fundamental fermions but the gauge bosons and the space-time metric (as well as the Higgs scalars, if any) are all composites of subquarks, the more fundamental and probably ultimate particles in nature. In this chapter, I shall review this supergrand unified subquark model, which is a candidate for the fundamental theory of composite particles and fields, based on the two principles (of relativity and of quantum) and the two hypotheses (of the fundamental length and of compositeness).

The theory effectively reproduces not only gauge theories for the strong and electroweak forces of the fundamental fermions but general relativity for gravity at low energies (or low temperature). At extremely high energies (or high temperature), it predicts an infinite series of nonlinear interactions of the fundamental fermions, the gauge bosons and the space-time metric. It also predicts the relations of the type

$$e^2 = 16\pi GM^2 \qquad (1)$$

where e, G and M are the gauge coupling constant, the Newtonian gravitational constant and the subquark mass. These relations indicate that the subquark mass may be as large as 10^{18}GeV.

2. The Two Principle and the Two Hypotheses

Let us start with introducing the two principles and the two hypotheses:

1) Relativity Principle

Among others, this requires that a theory be invariant under the Lorentz and general coordinate transformations. The velocity of a massless particle in the vacuum (which is the maximum speed), c ($\cong 3\times10^{10}$ cm/s), is set equal to unity for convenience as usual.

2) Quantum Principle

Among others, this requires that the theory be of quantum and that fields be quantized canonically. The Planck constant divided by 2π (a half of which is the minimum action), $\hbar$ ($\cong 1\times10^{-27}$ erg·s), is set equal to unity for convenience as usual.

3) Fundamental-Length Hypothesis

This requires that either one of the above two principles and the microscopic causality may break down if the distance between two particles becomes smaller than the fundamental length, ℓ, which may be as small as the Planck length, $G^{1/2}$ ($\cong 1\times10^{-33}$ cm).

4) Compositeness Hypothesis

This asserts that not only the fundamental fermions (leptons and quarks) but the gauge bosons and the space-time metric (as well as the Higgs scalars, if any) are all composites of subquarks, the more fundamental and probably ultimate particles in nature.

3. The Simplest Possible Model

Now I am ready to present the simplest possible model Lagrangian that satisfies these four conditions:

$$\begin{aligned} L' = {} & \bar{w}[e^{k\mu}\gamma_k(iD_\mu+\tau^i A^i_\mu)-\phi]w \\ & +g^{\mu\nu}(\partial_\mu-i\lambda^a G^a_\mu)C^\dagger(\partial_\nu-i\lambda^b G^b_\nu)C-\phi^2 C^\dagger C \\ & +\bar{f}wC+C^\dagger\bar{w}f \end{aligned} \tag{2}$$

where w and C are, respectively, the spinor and scalar subquarks ("wakems" and "chroms") which are the most fundamental particles. The auxiliary fields of f, A^i_μ, G^a_μ, ϕ, $e^{k\mu}$ and $g^{\mu\nu}$ are, respectively, the fundamental fermions, the gauge fields, the Higgs scalar, the vierbein and the space-time metric which represent various composite states (or collective motions) of subquarks. For definiteness and simplicity, the spinor and scalar subquarks are, respectively, assumed to form a fundamental N_w-plet of SU(N_w) and a fundamental N_C-plet of SU(N_C) (whose matrices are denoted by τ^i and λ^a for $i=1,2,3,\cdots,N_w^2-1$ and $a=1,2,3,\cdots,N_C^2-1$). Notice that the model Lagrangian is invariant not only under gauge transformation but under scale transformation.

4. The Effective Lagrangian

Define the effective Lagrangian for composite fields by the path- integral over the subquark fields:

$$\exp[i\int d^4x\sqrt{-g}L_{eff}] = \int dwd\bar{w}dCdC^{\dagger}\exp[i\int d^4x\sqrt{-g}L'] \tag{3}$$

where $g = \det g_{\mu\nu}$. The path-integration can be formally performed to give

$$\int d^4x\sqrt{-g}L_{eff} = -itr\ell n[e^{k\mu}\gamma_k(iD_\mu+\tau^i A^i_\mu)-\phi]$$

$$+itr\ell n[-(\sqrt{-g})^{-1}(\partial_\mu-i\lambda^a G^a_\mu)\sqrt{-g}g^{\mu\nu}(\partial_\nu-i\lambda^b G^b_\nu)-\phi^2]$$

$$-itr\ell n\{1+\bar{f}[e^{k\mu}\gamma_k(iD_\mu+\tau^i A^i_\mu)-\phi]^{-1}f[(\sqrt{-g})^{-1}(\partial_\nu-i\lambda^a G^a_\nu)\sqrt{-g}g^{\nu\kappa}(\partial_\kappa-i\lambda^b G^b_\kappa)$$

$$+\phi^2]^{-1}\} . \tag{4}$$

For further simplicity, the Higgs scalar ϕ will be replaced by its vacuum expectation value M which can be identified with the subquark mass and which is arbitrary because of the scale invariance. Then, expand the effective Lagrangian into an infinite series in the inverse power of M to obtain

$$\int d^4x\sqrt{-g}L_{eff} = i\sum_{n=1}^{\infty}\frac{1}{nM^n}tr[e^{k\mu}\gamma_k(iD_\mu+\tau^i A^i_\mu)]^n$$

$$+i\sum_{n=1}^{\infty}\frac{(-1)^{n-1}}{nM^{2n}}tr[(\sqrt{-g})^{-1}(\partial_\mu-i\lambda^a G^a_\mu)\sqrt{-g}g^{\mu\nu}(\partial_\nu-i\lambda^b G^b_\nu)]^n$$

$$+i\sum_{n=1}^{\infty}\sum_{\ell=0}^{\infty}\sum_{m=0}^{\infty}\frac{(-1)^m}{nM^{\ell+2m+3n}}$$

$$\times tr\{\bar{f}[e^{k\mu}\gamma_k(iD_\mu+\tau^i A^i_\mu)]^\ell f[(\sqrt{-g})^{-1}(\partial_\nu-i\lambda^a G^a_\nu)\sqrt{-g}g^{\nu\kappa}(\partial_\kappa-i\lambda^b G^b_\kappa)]^m\}^n . \tag{5}$$

Every term in this infinite series would be divergent and, therefore, ill-defined if there were no regularization in nature. According to the fundamental-length hypothesis, however, it must be finite. Suppose that invariance under both gauge and general coordinate transformations is kept tight even near the fundamental length. Then, I shall introduce a sofisticated way of regularization as illustrated as

$$tr\,[(\sqrt{-g})^{-1}\partial_\mu\sqrt{-g}g^{\mu\nu}\partial_\nu]$$

$$= \lim_{x'\to x}\int d^4x d^4y (\sqrt{-g_y})^{-1}\partial_{y\mu}\delta^4(y-x')[\exp(\int_x^y \Gamma^\alpha_{\beta\gamma}dz^\gamma)]^{\mu\nu}\sqrt{-g_x}\partial_{x\nu}\delta^4(x-y)$$

$$= -\frac{\delta^4(0)}{\sqrt{-g_0}}\int d^4x\sqrt{-g}R \tag{6}$$

and

$$\mathrm{tr}[(\partial_\mu + A_\mu)(\partial_\nu + A_\nu)]$$

$$= \lim_{x'\to x}\int d^4x d^4y\, \mathrm{tr}[(\partial_{y\mu}-A_{y\mu})\delta^4(y-x')\exp(-\int_x^y A_\alpha dz^\alpha)(\partial_{x\nu}+A_{x\nu})\delta^4(x-y)]$$

$$= \delta^4(0)\int d^4x\, \mathrm{tr}(\partial_\mu A_\nu - \partial_\nu A_\mu + [A_\mu, A_\nu]) \tag{7}$$

where $A_\mu = -i\tau^i A^i_\mu$, $\Gamma^\alpha_{\beta\gamma}$ and R are the Christoffel symbol and the Riemann scalar curvature, respectively, and $\delta^4(0)$ and g_0 are their values at the fundamental length.

More precisely, our regularization at the fundamental length consists of the following procedures: 1) Introduce the point-splitting in the trace operation with respect to the space-time coordinates as A(x)B(x) is replaced by $\lim_{x'\to x} A(x')B(x)$. 2) Insert the exponential factors such as $[\exp(\int_x^y \Gamma^\alpha_{\beta\gamma}dz^\gamma)]^{\mu\nu}$ and $\exp(-\int_x^y A_\alpha dz^\alpha)$ between the operator products so that they may be kept invariant under gauge and general coordinate transformations. 3) Perform the integration over the space-time coordinates in the trace operations. Then, there will appear the factor $\delta^4(0)$ defined by $\lim_{x'\to x}\delta^4(x'-x)$ which would be a divergent quantity in a theory without the fundamental length. 4) Take the factor $\delta^4(0)$ as a finite quantity of order ℓ^{-4}, according to the fundamental length hypothesis of $|x'-x| \gtrsim \ell$.

A careful calculation leads to the following simple result for the first several terms in the infinite series:

$$\int d^4x\sqrt{-g}L_{eff} = \frac{i\delta^4(0)}{\sqrt{-g_0}}\int d^4x\sqrt{-g}\,[\frac{2N_w+N_C}{M^2}R$$
$$- \frac{1}{8M^4}g^{\mu\nu}g^{\kappa\lambda}(4N_w F^i_{\mu\kappa}F^i_{\nu\lambda} + N_C G^a_{\mu\kappa}G^a_{\nu\lambda})$$
$$+ \frac{1}{M^4}\bar{f}e^{k\mu}\gamma_k(iD_\mu + \tau^i A^i_\mu + \lambda^a G^a_\mu)f + \cdots] \tag{8}$$

with

$$F^i_{\mu\nu} = \partial_\mu A^i_\nu - \partial_\nu A^i_\mu + \varepsilon^{ijk} A^j_\mu A^k_\nu$$

and (9)

$$G^a_{\mu\nu} = \partial_\mu G^a_\nu - \partial_\nu G^a_\mu + f^{abc} G^b_\mu G^c_\nu$$

where ε^{ijk} and f^{abc} are the structure constants of $SU(N_w)$ and $SU(N_C)$, respectively.

5. The Fundamental Relations

At low energies, this effective Lagrangian reproduces not only gauge theories for the strong and electroweak interactions of the fundamental fermions but general relativity for gravity if the Newtonian gravitational constant and the $SU(N_w)$ and $SU(N_C)$ gauge coupling constants, g and f, are identified as

$$\frac{i\delta^4(0)}{\sqrt{-g_0}} \frac{2N_w+N_C}{M^2} = \frac{1}{16\pi G}, \tag{10}$$

$$\frac{i\delta^4(0)}{\sqrt{-g_0}} \frac{N_w}{2M^4} = \frac{1}{4g^2} \quad \text{and} \quad \frac{i\delta^4(0)}{\sqrt{-g_0}} \frac{N_C}{8M^4} = \frac{1}{4f^2}. \tag{11}$$

These two identifications lead to the fundamental relations

$$g^2 = f^2 \frac{N_C}{4N_w} = 16\pi G M^2 \left(1 + \frac{N_C}{2N_w}\right), \tag{12}$$

which indicate that the subquark mass is as large as 10^{18}GeV.

They also indicate that the space-time at the fundamental length must be space-like, i.e. $g_0>0$, so that $i\delta^4(0)/\sqrt{-g_0} > 0$. Furthermore, since it is natural to suppose $i\delta^4(0)/\sqrt{-g_0}$ is of the order of ℓ^{-4}, the fundamental length ℓ must be of the order of $(16\sqrt{2}\pi G/e)^{1/2}$ ($\cong 10^{-18}$ GeV^{-1}). Notice that this result is similar to the relation in generalized Kaluza-Klein theories, $\ell = 2\pi\times (16\pi G/e^2)^{1/2}$ ($\cong 10^{-17}$GeV^{-1}) for the size ℓ of the extra dimensions, which has been carefully discussed by Freund[9] although the physical features involved are clearly different.

This theory also predicts at extremely high energies an infinite series of nonlinear interactions of the fundamental fermions, the gauge bosons and the space-time metric, which can be seen in the expression (5). The possible phase transition between the geometric phase (where $g_{\mu\nu} \neq 0$ and $|g_{\mu\nu}|<\infty$) and the pregeometric one (where $g_{\mu\nu} = \infty$) which Akama and myself[10] have discussed to present a possible explanation for the origin of the big bang of our Universe is one of the consequences of such nonlinear interactions at high temperature. There seem to be many other consequences, which are subjects for future investigations.

6. Conclusion

In conclusion, I would like to emphasize that the possibility of superheavy subquarks as "maximons",[11] which are called by Markov for the heaviest possible elementary particles in nature whose masses are of order of the Planck mass ($G^{-1/2} = 1\times10^{19}$GeV), seems to be one of the most theoretically attractive and noble ones although the possibility of relatively light subquarks whose masses are of order of the Fermi mass ($G_F^{-1/2}$ = 300Gev where G_F is the Fermi weak coupling constant) seems to be the most experimentally attractive and exciting one that may explain all the anomalous events recently found by the UA1 and UA2 groups in the CERN SPS $\bar{p}p$ Collider experiments.[8]

Note added:

The content of this chapter is the same as what I have just presented at the Third Seminar "Quantum Gravity" in Moscow.[12]

References and Footnotes

1. See for example, A. Einstein, The Meaning of Relativity, 3rd ed. (Princeton Univ. Press, Princeton, N.J., 1950), Appendix II; H. Weyl, Raum, Zeit und Materie (Springer Verlag, Berlin, Heidelberg, N.Y., 1920); T. Kaluza, Sitzker. Preuss. Akad. Wiss. Kl, 966 (1921); O. Klein, Z. Physik 37, 895 (1926).
2. See for example, W. Heisenberg, Z. Naturforsch. 14, 441 (1959), Introduction to the Elementary Particles (John Wiley & Sons, Ltd., London, N.Y., Sydney, 1966) and the earlier references therein; D. Ivanenko, Sov. Phys. 13, 141 (1938) and in Centenario di Einstein (Giunti-Barbara, Firenze, 1979), p.131 [Albert Einstein 1879-1979 (Johnston Reprint Corp., N.Y., 1979), p.295]; Y. Nambu and G. Jona-Lasinio, Phys. Rev. 96, 345 (1961).
3. See for example, J. Schwinger, Ann. Phys. (N.Y.) 2, 407 (1957); S.L. Glashow, Nucl. Phys. 22, 579 (1961); A. Salam, in Elementary Particle Physics, edited by N. Svartholm (Almqvist and Wiksell, Stockholm, 1968), p.367; S. Weinberg, Phys. Rev. Lett. 19, 1264 (1967).
4. See for example, J.C. Pati and A. Salam, Phys. Rev. D10, 275 (1974); H. Georgi and S.L. Glashow, Phys. Rev. Lett. 32, 438 (1974).
5. H. Terazawa, Phys. Rev. D22, 184 (1980). See also H. Terazawa, Y. Chikashige and K. Akama, Phys. Rev. D15, 480

(1977); H. Terazawa, Y. Chikashige, K. Akama and T. Matsuki, Phys. Rev. D15, 1181 (1977); H. Terazawa, Phys. Rev. D16, 2373 (1977), D22, 1037 (1980) and D22, 2921 (1980); H. Terazawa and K. Akama, Phys. Lett. 96B, 276 (1980) and 97B, 81 (1980); H. Terazawa, Phys. Lett. 101B, 43 (1981) and 133B, 57 (1983). For reviews, see for example, H. Terazawa, in Proc. INS Int. Symp. on New Particles and the Structure of Hadrons, Tokyo, July 12-14, 1977 and France-Japan Joint-Seminar on New Particles and Neutral Currents, Tokyo and Kyoto, July 14-16 and 18, 1977, edited by K. Fujikawa et al. (INS, Univ. of Tokyo, Tokyo, 1978), p.231 and p.579; in Proc. XIX Int. Conf. on High Energy Physics, Tokyo, Aug., 23-30, 1978, edited by S. Homma et al. (Phys. Soc. Japan, Tokyo, 1979), p.617, in Proc. Second Marcel Grossmann Meeting on the Recent Developments of General Relativity, ICTP, Trieste, 5-11 July, 1979, edited by R. Ruffini (North-Holland Pub. Co., Amsterdam, New York, Oxford, 1982), p.519; in Proc. Japan-Italy Symposium on Fundamental Physics, Tokyo, Jan. 27-30, 1981, edited by S. Fukui and T. Toyoda (Nagoya Univ., Nagoya, 1981), p.65; in Proc. 1981 INS Symp. on Quark and Lepton Physics, Tokyo, June 25-27, 1981, edited by K. Fujikawa et al. (INS, Univ. of Tokyo, Tokyo, 1981), p.296; in Proc. XXI Int. Conf. on High Energy Physics, Paris, July 26-31, 1982, edited by P. Petiau and M. Porneuf, J. de Phys. C3-289 (1982); in Proc. Third Marcel Grossmann Meeting on the Recent Developments of General Relativity, Shanghai, Aug. 30-Sept. 3, 1982, edited by Hu Ning (Science Press and North-Holland Pub. Co., Beijing, 1983), p.239; in Proc. Topical Symposium on High Energy Physics, Tokyo, Sept. 7-11, 1982, edited by T. Eguchi and Y. Yamaguchi (World Scientific Pub. Co., Singapore, 1983), p.173; in Proc. Europhysics Conf. on Flavor Mixing in Weak Interactions, Erice, March 4-11, 1984, edited by L.-L. Chau (Plenum Pub. Co., New York, 1984), p.655; in Proc. XXII Int. Conf. on High Energy Physics, Leipzig, July 19-25, 1984, edited by A. Meyer and E. Wieczorek (Akademie der Wissenschaften der DDR, Zeuthen, 1984), Vol.I, p.63.

6. J. D. Bjorken, Ann. Phys. (N.Y.) 24, 174 (1963); H. Terazawa, Y. Chikashige and K. Akama, in Ref. 5; H. Terazawa and K. Akama, in Ref. 5; H. Terazawa, in Ref. 5.
7. A.D. Sakharov, Doklady Akad. Nauk SSSR 177, 70 (1967) [Sov. Phys. JETP 12, 1040 (1968)] and Teor. Mat. Fiz. 23, 178 (1975) [English translation (Plenum Pub. Co., New York, 1979), p.435]; H. Terazawa, Y. Chikashige, K. Akama and T. Matsuki, in Ref. 5; K. Akama, Y. Chikashige, T. Matsuki and H. Terazawa, Prog. Theor. Phys. 60, 868 (1978); H. Terazawa and K. Akama, in Ref. 5; H. Terazawa, in Ref. 5; K. Akama and H. Terazawa, Gen. Rel. Gravit. 15, 201 (1983). For reviews, see for example, H. Terazawa, in Proc. Second Seminar "Quantum Gravity", Academy of Sciences of the USSR, Moscow, Oct. 13-15, 1981, edited by M.A. Markow and P.C. West (Plenum Pub. Co., London, New York, 1984), p.47 and reviews in Ref. 5. See also S.L. Adler, Rev. Mod. Phys. 54, 729 (1982).

8. For the latest review, see H. Terazawa, in Ref. 5.
9. P.G.O. Freund, Phys. Lett. 120B, 335 (1983). See also Y.M. Cho and P.G.O. Freund, Phys. Rev. D12, 1711 (1975).
10. K. Akama and H. Terazawa, in Ref. 7. For reviews, see H. Terazawa, reviews in Refs. 5 and 7.
11. M.A. Markov, Prog. Theor. Phys. Suppl., Extra Number, 85 (1965) and ZhETF 51, 878 (1966) Sov. Phys. JETP 24, 584 (1967) .
12. H. Terazawa, to be published in Proc. Third Seminar "Quantum Gravity", Academy of Sciences of the USSR, Moscow, Oct. 23-25, 1984, edited by M.A. Markov and P.C. West (World Scientific Pub. Co., Singapore, 1985).

IV. CONCLUDING REMARK

Finally, let me conclude these lectures by addressing the following few words to Mexican youngsters in my poor Spanish:

¡SENOR, SENORA Y SENORITA EN MEXICO!

¡NO PARE AQUI, POR FAVOR!

¡DESE PRISA, POR FAVOR!

¿PUEDE USTED VER POPO?

MUCHAS GRACIAS.

¡ADIOS!

H. TERAZAWA

14 DICIEMBRE 1984

OAXTEPEC, MORELOS, MEXICO.

NEUTRINO MASSES*

D. Wyler
Theoretische Physik, ETH, 8093 Zürich (Switzerland)

ABSTRACT

I give an overview on theoretical aspects of neutrino masses and the experimental methods and results.

1. INTRODUCTION

The neutrino was introduced 1930 by Pauli[1] as a last resort to save energy-momentum conservation in β-decay. Since then it has been established as one of the fundamental constituents of matter, on the same footing with the electron, the up and the down quark. With the discovery of more "flavors" of quarks and charged leptons, also more neutrinos were found. At present, there are three neutrinos, ν_e, ν_μ, ν_τ, whose masses are bound from above as follows[2]

$$\begin{aligned} m(\nu_e) &< 40 \quad \text{eV} \\ m(\nu_\mu) &< 260 \quad \text{keV} \\ m(\nu_\tau) &< 160 \quad \text{MeV} \end{aligned} \tag{1.1}$$

Whereas ν_e, ν_μ have been experimentally shown to be distinct, the fact that ν_τ is a new state is inferred indirectly from anomaly cancellations (theoretically) or from absence of lepton flavor changing processes, such as $\tau \to \tau\mu$, $\tau \to \pi e$.**

We see from (1.1) that the neutrinos are substantially lighter than the corresponding leptons, in particular ν_e. Whether the mass of the neutrinos is really zero or just very small, there might be a new physical principle responsible for it beyond the standard model***, and the neutrinos might show the first sign for it (of

* Lectures given at the Escuela Mexicana de Particulas y Campos; Oaxtepec, Mexico, December 3-14, 1984.

** The term neutrino sometimes also applies to heavy neutral leptons, introduced in extended models. We will use it for both light (massless) and heavy neutral leptons.

*** It should be noted that the values of the other fermion's masses are equally not understood within the standard model.

0094-243X/86/1430171-90$3.00

course, there can be "new" physics which has no special bearing on neutrino masses, such as supersymmetry[3]).

The theoretical picture might be summarized as follows. Models with an SU(2) x U(1) or SU(5) [4] structure tend to have vanishing neutrino mass or a "majorana" mass accompanied by a massless (or very light) boson. Such models have a somewhat unpleasant asymmetry between neutrinos and the other fermions. Models based on a left-right symmetric theory have necessarily a right-handed neutrino; they lead usually to a very heavy and a very light majorana neutrino, or to a massless and a relatively heavy Dirac neutrino*. Due to the usual lack of predictability, the values for the masses are not known, but they are expected to be 10^{-5} - 1eV for the light and 10^{9} - 10^{15} GeV for the heavy neutrino in the first case and few GeV - 10^{4} GeV in the second. These values reflect, apart from the unknown coupling constants, expected mass scales of new physics, mainly the GUT scale and the mass scale of the right handed gauge bosons. Although there exist so far no analysis of neutrino masses in "deeper" theories, such as technicolor[5] or models of composite quarks and leptons[6]; one might speculate that, particularly in the latter, considerations similar to those of left-right models hold. This, because if it is possible to "assemble" a right-handed electron one expects to be able to do so also for the neutrino[7]. Which of all these possibilities we should choose, is, of course, open.

The expected smallness of the neutrino mass has led to many very interesting experimental techniques, which are able to measure various aspects of the mass with increasing precision. Most experiments give upper bounds; two seem to indicate non-zero masses. However, there are still uncertainties in the systematics and one will have to wait for a more conclusive picture. Unfortunately, some of the small values predicted (see above) might always stay outside the reach of experiment.

A powerful tool to investigate neutrino masses are the developments in astrophysics and cosmology. Due to the picture of the hot early universe, light neutrinos are very abundant in the universe (because of their weak interactions they disappear extremely slowly). Therefore, even a very small neutrino mass can make the total neutrino mass the most important in the universe, with decisive consequences on the development of the universe.

In these lectures I want to treat these various aspects of the neutrino mass. I have attempted to describe various topics in some detail and to be as complete as possible in enumerating and referencing relevant items.

---

* Also in SU(2) x U(1) or SU(5) one can introduce a right-handed neutrino; however, it is not required, in contrast to the other fields.

2. TECHNICAL DETAILS

a) Transformation properties[8]

To describe the various mass terms, we consider various forms of spinors and their transformation properties under certain symmetries.

Let ψ denote a complex 4-spinor, a Dirac spinor. It describes a particle, its antiparticle, both with two helicities. We define:

Helicity states: $\psi_L \equiv \frac{1}{2}(1-\gamma_5)\psi \equiv L\psi$ left-handed (2.1.1)

$\psi_R \equiv \frac{1}{2}(1+\gamma_5)\psi \equiv R\psi$ right-handed (2.1.2)

In an SL(2,C) notation for the Lorentz group, ψ_L transforms as (1/2,0) , and ψ_R as (0,1/2) .

Charge conjugation (antiparticle):

$$\psi_C \equiv C\bar{\psi}^T = C\gamma_0^T\psi^* \tag{2.2}$$

$$\bar{\psi} = \psi^+\gamma_0$$

$$\begin{aligned} C\gamma_\mu^T C^{-1} &= -\gamma_\mu \\ CC^+ &= 1 \\ C^T &= -C \\ C\gamma_5 &= \gamma_5^T C \\ \gamma_5\psi_C &= -(\gamma_5\psi)_C \end{aligned} \tag{2.3}$$

(This form of ψ_C is obtained when considering the Dirac equation with the photon field and attempting to find a spinor with opposite charge.) In the Bjorken-Drell[8] notation

$$C = i\gamma^0\gamma^2 = \begin{pmatrix} 0 & \begin{smallmatrix} 0 & 1 \\ -1 & 0 \end{smallmatrix} \\ \begin{smallmatrix} 0 & 1 \\ -1 & 0 \end{smallmatrix} & 0 \end{pmatrix} \tag{2.4}$$

Furthermore, the parity transform is[8]

$$\psi_P = \eta_P\gamma_0\psi \qquad \eta_P = \pm 1 \tag{2.5}$$

In gauge theories we generally consider the states with definite helicity - in fact, the neutrinos seem to have only a left-handed component. We can then calculate $(\psi_L)_C$, $(\psi_R)_C$ for such spinors:

$$\begin{aligned} (\psi_L)_c &= (L\psi)_c = C\gamma_0^T L^* \psi^* = R\psi_c \\ (\psi_R)_c &= (R\psi)_c = C\gamma_0^T R^* \psi^* = L\psi_c \end{aligned} \tag{2.6}$$

$$\begin{aligned} (\psi_L)_P &= (L\psi)_P = \eta_P \gamma_0 L \psi = R\eta_P \psi_P \\ (\psi_R)_R &= (R\psi)_P = \eta_P \gamma_0 R \psi = L\eta_P \psi_P \end{aligned} \tag{2.7}$$

Thus the charge conjugated state of a left-handed (right-handed) spinor is a right-handed (left-handed) spinor. Since, by CPT, there always exists the charge-conjugated particles, all left (right) handed states can be written as right (left) handed states with opposite quantum numbers. We will denote:

$$(\psi_L)_c \text{ by } \psi_{RC} \qquad (\psi_R)_c \text{ by } \psi_{LC} \tag{2.8}$$

b) Majorana fields[9]

If a fermion is neutral, one can define a spinor which is self conjugate, e.g.

$$\psi_C = a\psi \qquad |a| = 1 \tag{2.9}$$

This is the Majorana condition. It is possible to define (2.9), because with ψ , also ψ_c satisfies the Dirac equation and thus any linear combination. We define a Majorana-spinor, χ , from a Dirac spinor as follows

$$\chi = \frac{1}{\sqrt{2}}(\psi + \eta\psi_c) \qquad \chi_c = \eta^* \chi \tag{2.10}$$

The Majorana-spinor has half as many degrees of freedom as a Dirac spinor, namely 4. Thus, written as a 4-spinor, it can have real components. The corresponding representation of the γ matrices is

$$\gamma^0 = \begin{pmatrix} 0 & \sigma^2 \\ \sigma^2 & 0 \end{pmatrix} \quad \gamma^1 = \begin{pmatrix} i\sigma_3 & 0 \\ 0 & i\sigma_3 \end{pmatrix} \quad \gamma^2 = \begin{pmatrix} 0 & -\sigma^2 \\ \sigma^2 & 0 \end{pmatrix} \quad \gamma^3 = \begin{pmatrix} -i\sigma_1 & 0 \\ 0 & -i\sigma_1 \end{pmatrix}$$

If we write $\psi = \psi_L + \psi_R$, then the helicity states of χ are

$$\chi_L = \frac{1}{\sqrt{2}}\left(\psi_L + \eta(\psi_R)_c\right)$$
$$\chi_R = \frac{1}{\sqrt{2}}\left(\psi_R + \eta(\psi_L)_c\right) \qquad (2.11)$$

Under the parity transformation (2.5) we have

$$\chi_P = \frac{1}{\sqrt{2}}\left(\psi_P + \eta(\psi_c)_P\right) = \frac{1}{\sqrt{2}}\left(\eta_P\gamma_0\psi + \eta C\gamma_0^T\eta_P^*\gamma_0^*\psi^*\right)$$
$$= \frac{1}{\sqrt{2}}\left(\eta_P\psi - \eta_P^*\eta\psi_c\right) \qquad (2.12)$$

(using $\gamma_0^+ = \gamma_0$) .
Only, if $\eta_P = \pm i$, χ is an eigenstate of parity:

$$\chi_P = \pm i\gamma_0\chi \qquad (2.13)$$

Thus, the parity-phase of a Majorana field is purely imaginary.[10]

c) Mass terms

We consider the Dirac equation for a free field an the Lagrangian which leads to it. Usually, for Dirac spinors we write

$$\mathcal{L} = i\bar{\psi}\not{\partial}\psi - m\bar{\psi}\psi \qquad (2.14)$$

The second, or mass term, has the form $(\bar{\psi}_L\psi_R + \bar{\psi}_R\psi_L)$. More generally, any mass term must be of the form $\bar{\psi}_L\phi_R$, $\bar{\psi}_R\phi_L$: because $\phi_L \sim (\frac{1}{2},0)$, $\bar{\psi}_R = (\psi_R)_C^T\ C^* \sim (\frac{1}{2},0)$ etc., these are the only Lorentz invariant terms one can construct. In the following we discuss the mass terms which can be formed from ψ and ψ_C ; the existence of ψ_C is a consequence of the CPT-theorem.

c.1) Non-chiral fields

The ordinary mass term, or Dirac mass term, is of the form (ψ = Dirac spinor)

$$\mathcal{L}_M = m_D \bar{\psi}\psi \tag{2.15}$$

For several fields ψ, m_D is a matrix. Because of

$$(\bar{\psi}_1 \psi_2)^+ = (\psi_2^+ \gamma_0^+ \psi_1) = \bar{\psi}_2 \psi_1 \tag{2.16}$$

m_D is hermitean. Also, we have

$$\begin{aligned} \bar{\psi}_C^1 \psi_C^2 &= \psi_C^{1+} \gamma_0 \psi_C^2 = \psi^{1T} \gamma_0^* C^+ \gamma_0 C \gamma_0^T \psi^{*2} \\ &= -\psi^{1T} \gamma_0^* \psi^{*2} = \psi^{+2} \gamma_0^+ \psi^1 \\ &= \bar{\psi}^2 \psi^1 \qquad \uparrow \{\psi^1, \psi^{2*}\} = 0 \end{aligned} \tag{2.17}$$

We can now write

$$\mathcal{L}_M = \frac{m_D}{2}(\bar{\psi}\psi + \bar{\psi}_C \psi_C) = \frac{m_D}{2}(\bar{\chi}_+ \chi_+ + \bar{\chi}_- \chi_-) \tag{2.18}$$

where

$$\chi_\pm = \frac{1}{\sqrt{2}}(\psi \pm \psi_C) \quad (\chi_\pm)_C = \pm \chi_\pm \tag{2.19}$$

We see clearly that the Dirac spinor describes the particles ψ, ψ_C; also the Dirac spinor is equivalent to two Majorana spinors of opposite C-parity and the same mass; this is like writing the charged $W^\pm$ as two real fields W^1, W^2.

Besides $\bar{\psi}\psi$ we can also form scalars $\bar{\psi}\psi_C$ or $\bar{\psi}_C\psi$. We write more general than (2.19)

$$\mathcal{L}_M = \frac{m_D}{2}(\bar{\psi}\psi + \bar{\psi}_C \psi_C) + \frac{1}{2}(m_1 \bar{\psi}\psi_C + m_2 \bar{\psi}_C \psi) \tag{2.20}$$

Because of

$$(\bar{\psi}^1 \psi_C^2)^+ = (\psi^{2+}{}_C \gamma_0^+ \psi^1) = (\bar{\psi}_C^2 \psi^1) \tag{2.21}$$

it follows that $m_2 = m_1$ and that (2.20) is hermitean. Also, note that

$$\bar{\Psi}_c^1 \psi^2 = \psi_c^{1+} \gamma_0 \psi^2 = -\psi^{2T} \gamma_0^T \psi_c^{1*} = \tag{2.22}$$
$$= \psi^{2T} C^+ \gamma_0 C \psi_c^{1*} = \bar{\Psi}_c^2 \gamma_0 C \gamma_0^* C^+ \gamma_0 C C^* \gamma_0^+ \psi^1$$
$$= \bar{\Psi}_c^2 \psi^1$$

This means, that if $m_2 \to$ matrix, then it is symmetric; Furthermore, (2.21) implies that $m_1 = m_2^*$. With (2.17) and (2.21) we can write the multi-flavor version of (2.20) as

$$\mathcal{L}_M = \frac{1}{2}(\bar{\Psi}, \bar{\Psi}_c) \begin{pmatrix} M_D & M_M \\ M_M^* & M_D^T \end{pmatrix} \begin{pmatrix} \psi \\ \psi_c \end{pmatrix} \tag{2.23}$$

with an hermitean matrix $(M_M = M_M^T)$

$$M = \frac{1}{2} \begin{pmatrix} M_D & M_M \\ M_M^* & M_D^T \end{pmatrix} \tag{2.24}$$

M_D is the Dirac term, M_M is the Majorana term. We can define a fermion number charge through

$$\psi \to e^{i\alpha} \psi$$
$$\psi^* \to e^{-i\alpha} \psi^* \tag{2.25}$$

M_D is invariant under this transformation, M_M , however, violates it by two units. It reflects the fact that for Majorana fermions, fermionic charge is not a good quantum number, as well as the other quantum numbers carried by ψ . M can be diagonalized by a unitary matrix:

$$U^{+} M U = \begin{pmatrix} m_1 & & \\ & m_2 & \\ & & \ddots \end{pmatrix} \tag{2.26}$$

m_i positive, real. (2.24) implies that[10]

$$U = \begin{pmatrix} A & -B \\ A^* & B^* \end{pmatrix} \tag{2.27}$$

and then

$$\phi_+ = A^+ \psi + A^T \psi_c \qquad (\phi_\pm)_c = \pm \phi_\pm$$
$$\phi_- = -B \psi + B^T \psi_c$$

are the mass eigenstates. The eigenvalues are the elements of the diagonal matrices

$$Re\,(A^+ M_D A + A^+ M_M A^*) \tag{2.28}$$

$$Re\,(B^+ M_D B - B^+ M_M B^*) \tag{2.29}$$

For the trivial case (2.18), $A = \frac{1}{\sqrt{2}} = B$ and (2.19) are recovered. For the 2 x 2 matrix M

$$M = \begin{pmatrix} m & \tilde{m} e^{i\alpha} \\ \tilde{m} e^{-i\alpha} & m \end{pmatrix} \tag{2.30}$$

$A = B = \frac{1}{\sqrt{2}} e^{i\alpha/2}$; the eigenvalues and eigenvectors are

$$\lambda_\pm = m \pm \tilde{m}$$
$$\phi_\pm = \frac{1}{\sqrt{2}} (e^{-i\alpha} \psi \pm e^{i\alpha} \psi_c) \tag{2.31}$$

The general result is, that for n fields ψ , there are 2n Majorana fields; n which have C-parity $+\eta_i$, n which have C-parity $-\eta_i$.

Consider finally, the CP properties. Using (2.12), (2.13) one shows directly (η = C-phase of ψ)

$$
\begin{aligned}
\psi_{CP} = (\psi_P)_C &= \pm \eta \gamma_0 \psi_c \\
(\psi_c)_{CP} &= \mp \eta^* \gamma_0 \psi \\
(\bar{\psi})_{CP} &= \pm \bar{\psi}_c \eta^* \gamma_0 \\
(\psi_{CP})_{CP} &= -(\eta^*)^2 \psi \\
(\bar{\psi}\psi_c)_{CP} &= -(\eta^*)^2 \bar{\psi}_c \psi_c
\end{aligned}
\tag{2.32}
$$

If CP is to be conserved, then the last relation implies that $M_M = (-1)(\eta^*)^2 M_M^*$; also since the mass eigenstates are CP eigenstates, we see from the first two relations $\eta^* = -\eta$, because only then the linear combination $\psi + \alpha\psi_c$ can be a CP eigenstate. Thus M_M is then real.

We can also determine the propagators. The total free Lagrangian is

$$
\mathcal{L} = (\bar{\psi}\ \psi_c)\ \frac{1}{2} \begin{pmatrix} \not{p} - M_D & -M_M \\ -M_M & \not{p} - M_D \end{pmatrix} \begin{pmatrix} \bar{\psi} \\ \psi \end{pmatrix}
\tag{2.33}
$$

The propagator is obtained as usual by inverting the matrix in (2.3), this gives then terms of the form $\langle\bar{\psi}\psi\rangle$, $\langle\bar{\psi}_c\psi_c\rangle$, but also

$\langle\bar{\psi}\psi_c\rangle$ and $\langle\bar{\psi}_c\psi\rangle$, which can also be written as $\langle\bar{\psi}C\bar{\psi}\rangle$ or $-\langle\psi^T C^{-1}\psi\rangle$. Instead of inverting (2.3) we diagonalize directly and get

$$\mathcal{L} = \frac{1}{2}\sum_i \bar{\chi}_i (\not{p} - m_i)\chi_i \qquad (\chi_i)_c = \eta_i \chi_i \tag{2.34}$$

This gives directly the propagator[11]

$$\langle\bar{\chi}_i\, \chi_i\rangle = \frac{i(\not{p} + m_i)}{p^2 - m_i^2} \tag{2.35}$$

but since $\chi_i^T = \eta^* \bar{\chi}_{ic} C = \bar{\chi}_i\, C^T$, we have also

$$\langle\chi_i^T \chi_i\rangle = \frac{iC(\not{p} + m_i)\eta}{p^2 - m_i^2} \tag{2.36}$$

The last term, although it looks entirely like a term which violates fermion number by two units, contain both, fermion number violating and non-violating terms.

If we consider states of definite chirality then

$$\langle\chi_L^T \chi_L\rangle = \frac{iCm}{p^2 - m^2} \tag{2.37}$$

$$\langle\chi_R^T \chi_L\rangle = \frac{iC\not{p}}{p^2 - m^2} \tag{2.38}$$

The first term vanishes for small m , the second does not. This may seemingly give rise to large fermion number violations for small m . However, when $m_M = 0$, it contains only fermion-number conserving pieces. Note, that if $\psi \equiv \psi_L$ only, then $\psi_C = \psi_{RC}$, and the only fermion number violation is in (2.37). Thus, the large effect of (2.38) only appears if there are ψ_R and ψ_L and there are Dirac and Majorana masses.

c.2 chiral fields

As pointed out previously, chiral fields are the natural way to describe the fermions in a gauge theory. We therefore want to reexpress the above in such spinors. Since a mass term is always of the form $\overline{\psi}_L \phi_R$ or $\overline{\psi}_R \phi_L$, the most general mass term is

$$\mathcal{L}_M = (\overline{\psi}_{RC} \; \overline{\psi}_R) M \begin{pmatrix} \psi_L \\ \psi_{LC} \end{pmatrix} + (\overline{\psi}_L \; \overline{\psi}_{LC}) M^+ \begin{pmatrix} \psi_{RC} \\ \psi_R \end{pmatrix} \qquad (2.39)$$

$$M = \begin{pmatrix} M_L & M_D \\ M_D^T & M_R \end{pmatrix} \qquad M^T = M \qquad (2.40)$$

where M_L, M_R are $n_L \times n_L$ and $n_R \times n_R$ matrices, representing Majorana masses, and M_D is a $n_L \times n_R$ matrix, the Dirac mass. n_L and n_R are generally different; n_L is the number of left-handed fields, n_R that of right-handed fields. In the standard model with three neutrinos, $n_L = 3$, $n_R = 0$. M_L, M_R are symmetric (see (2.21)), the fact that M_D^T enters is due to $\overline{\psi}_{RC}\, \psi_{LC}^j = \overline{\psi}_R^j\, \psi_L^j$. The symmetry and fermion number violating character of $M_{L,R}$ comes even better out if we set

$$\overline{\psi}^i_{RC}\, \psi_L^j = \psi_L^{iT} C^{-1} \psi_L^j \qquad (2.41)$$

Also, the Lorentz-structure is clearly visible this way: in the "chiral" basis in which we work, $\gamma_5 = \begin{pmatrix} 1 & 0 \\ 0 & -1 \end{pmatrix}$ is diagonal. Then $C^{-1} = \begin{pmatrix} i\sigma^2 & 0 \\ 0 & i\sigma^2 \end{pmatrix}$. Thus, in the left (right)-handed sectors, $(C^{-1})_{ab} = \epsilon_{ab}$, $\epsilon_{12} = -\epsilon_{21} = 1$, $\epsilon_{11} = \epsilon_{22} = 0$. Then (2.4) is clearly an $SU(2) \times SU(2)$ singlet, because $\psi \sim (\frac{1}{2},0)$ and the Clebsch-Gordan coefficients for $\frac{1}{2} \otimes \frac{1}{2} \to 0$ are just ϵ_{ab}. The antisymmetric combination is also in accord with Fermi-statistics.

The free Lagrangian has the form

$$\mathcal{L} = \frac{1}{2}\left(\bar{\psi}_L\ \bar{\psi}_{LC}\ \bar{\psi}_{RC}\ \bar{\psi}_R\right)\begin{pmatrix} \not{p} & 0 & M_L & M_D^* \\ 0 & \not{p} & M_D^+ & M_R^+ \\ M_L & M_D & \not{p} & 0 \\ M_D^T & M_R & 0 & \not{p} \end{pmatrix}\begin{pmatrix} \psi_L \\ \psi_{LC} \\ \psi_{RC} \\ \psi_R \end{pmatrix} \tag{2.42}$$

from which we can readily infer the propagators for the various fields (see 2.35 - 2.38).

As in the previous case we consider the minimal example, a 2 x 2 matrix

$$M = \begin{pmatrix} m_L & m_D \\ m_D & m_R \end{pmatrix} \tag{2.43}$$

M can be diagonalized by

$$U^+ M U = \begin{pmatrix} \lambda_1 & 0 \\ 0 & \lambda_2 \end{pmatrix}$$

$$\lambda_{1,2} = \frac{1}{2}\left\{(m_L + m_R) \pm \sqrt{(m_L - m_R)^2 + 4m_D^2}\right\} \tag{2.44}$$

$$U = \begin{pmatrix} \cos\alpha & -\sin\alpha \\ \sin\alpha & \cos\alpha \end{pmatrix};\ tg2\alpha = \frac{2m_D}{m_L - m_R} \tag{2.45}$$

(The previous case (2.30) corresponds to $m_L = m_R$). We now get two mass eigenstates:

$$\begin{aligned} \chi_1 &= (c\psi_L + s\psi_{LC}) + (c\psi_{RC} + s\psi_R) \\ \chi_2 &= (-s\psi_L + c\psi_{LC}) + (-s\psi_{RC} + c\psi_R) \\ & c, s = \cos\alpha, \sin\alpha \end{aligned} \tag{2.46}$$

and both have positive C-parity.

If m_D is large enough, λ_2 can become negative. Since we usually prefer positive masses, we can change the eigenvalue of X_2 by letting

$$\chi_2 \rightarrow \gamma_5 \chi_2$$

or

$$\chi_2 \rightarrow \chi_2' = (s\chi_L - c\chi_{LC}) + (-s\chi_{RC} + c\chi_R) \qquad (2.47)$$

Thus, the state X_2' with positive mass has a negative C-parity.* We note the special cases

$$m_D = 0 \qquad \chi_1 = \frac{1}{\sqrt{2}}(\chi_L + \chi_{RC}) \qquad \lambda_1 = m_L$$
$$\chi_2 = \frac{1}{\sqrt{2}}(\chi_R + \chi_{LC}) \qquad \lambda_2 = m_R$$

$$(2.48)$$

$$m_R = m_L = 0 \qquad \chi_1 = \frac{1}{\sqrt{2}}(\chi + \chi_c) \qquad \lambda_1 = m_D$$
$$\chi_2 = \frac{1}{\sqrt{2}}(\chi - \chi_c) \qquad \lambda_2 = m_D$$
$$\chi = \chi_L + \chi_R \ .$$

We see that only if both m_D, $m_M \neq 0$, the mass eigenstates contain particle and antiparticle with the same helicity.

This can be generalized. Since M is symmetric, one can always write

$$M = U^T M_{Di} U \qquad (2.49)$$

where M_{Di} is diagonal. In general, M_{Di} is not real (nor even positive). This is in contrast to the case (2.24) where M was hermitean. By a change in U, e.g.

---

* This result appears to be in contradiction with the comments after Eq. (2.31). There is no contradiction because the mass matrices are in different bases. If e.g. $m_L = m_R = 0$, (2.47) has $\lambda_1 = -\lambda_2$ while (2.31) leads to $\lambda_1 = \lambda_2$. (2.47) restores $\lambda_1 = \lambda_2$ but gives now the extra negative charge conjugation property.

$$\mathcal{U} \rightarrow \sqrt{S}\,\mathcal{U} \qquad \mathcal{U}^T \rightarrow \mathcal{U}^T\sqrt{S} \tag{2.50}$$

where S is a diagonal matrix of phases, we can write

$$M = \mathcal{U}^T \sqrt{S}\, |M_{Di}| \sqrt{S}\, \mathcal{U} \tag{2.51}$$

where $|M_{Di}|$ is diagonal with positive entries.*

Now, if we write

$$\chi = \mathcal{U} \begin{pmatrix} \psi_L \\ \psi_{LC} \end{pmatrix} + S^* \mathcal{U}^* \begin{pmatrix} \psi_{RC} \\ \psi_R \end{pmatrix} \tag{2.52}$$

then

$$\begin{aligned} \bar{\chi} |M_{Di}| \chi &= (\bar{\psi}_{RC}\ \bar{\psi}_R)\, \mathcal{U}^T S |M_{Di}| \mathcal{U} \begin{pmatrix} \psi_L \\ \psi_{LC} \end{pmatrix} \\ &\qquad + h.c. \\ &= (\bar{\psi}_{RC}\ \bar{\psi}_R)\, M \begin{pmatrix} \psi_L \\ \psi_{LC} \end{pmatrix} + h.c. \end{aligned} \tag{2.53}$$

since S commutes with $|M_{Di}|$. Also

$$\chi_c = S\chi \tag{2.54}$$

We notice that S and thus the C-parity of X depends on the choice of U. For example, if M is real, then $S_{ii} = \pm 1$, if U is taken to be a (real) orthogonal matrix. Then, if one of the $S_{ii} = -1$ (negative C-parity, see (2.54)) then by letting $U \rightarrow \begin{pmatrix} 1 & & \\ & i & \\ & & \ddots \end{pmatrix} \cdot U$, the C-parity changes, but since U has now phases, it might appear that we have a CP-violating theory, which is not the case.[12] We will consider the effects of $S_{ii} = -1$ later in double β-decay.

---

* The matrix S can be calculated from M and U using a result of Schur (see e.g. R. Slansky, Phys. Rep. 79 (1981) 1, also J. Bernabéu and P. Pascual, Ref. 12)

Examples

Consider the case $n_L = 2$, $n_R = 0$ where M_L has the special form (we call $\Psi_L^1 = \nu_e$, $\psi_L^2 = \nu_\mu$)

$$M_L = \begin{pmatrix} 0 & c \\ c & 0 \end{pmatrix} \tag{2.55}$$

Since

$$\mathcal{L}_M = \begin{pmatrix} \bar{\nu}_{ec} \\ \bar{\nu}_{\mu c} \end{pmatrix}^T \begin{pmatrix} 0 & c \\ c & 0 \end{pmatrix} \begin{pmatrix} \nu_e \\ \nu_\mu \end{pmatrix} + h.c. \tag{2.56}$$

is invariant under $\nu \to e^{i\alpha}\nu$, $\nu_\mu \to e^{-i\alpha}\nu_\mu$, ("e-number"-"μ-number") is conserved. (2.55) looks like a Dirac mass matrix[13], however, since e and μ are different, in higher order new effects can come, such as in the calculation of the magnetic moment (which vanishes for a Majorana particle)[14]

$$\mu_{Dirac} = \frac{3eG_F m_\nu}{8\pi^2\sqrt{2}} \quad \text{for a Dirac neutrino} \tag{2.57.1}$$

$$\mu_{D\text{-}M} = \frac{3eG_F C}{8\pi^2\sqrt{2}M_W^2}(m_\mu^2 - m_e^2) \quad \text{for (2.)} \tag{2.57.2}$$

where (2.57) are calculated with a one-loop graph with internal W , charged leptons.

If

$$M_L \to \begin{pmatrix} \delta & c \\ c & \delta \end{pmatrix} \qquad \delta \ll c \tag{2.58}$$

then the eigenstates are largely Dirac, they are refered to as "Pseudo-Dirac".[13]

Matrices of the type (2.56) lead to conserved quantum numbers. In general, if

$$M_L = \begin{pmatrix} 0 & \ast \\ \ast & 0 \end{pmatrix}\begin{matrix}\}n_A \\ \}n_B\end{matrix} \qquad \mathcal{L} = \begin{pmatrix} \bar{\psi}_{AC} \\ \bar{\psi}_{BC} \end{pmatrix}^T M_L \begin{pmatrix} \psi_A \\ \psi_B \end{pmatrix} + h.c. \tag{2.59}$$

then (number of particles A - number of particles B) is conserved. There are $|n_A - n_B|$ massless particles.[15]

Electromagnetic properties of Majorana particles[16]

Consider the expression

$$O_\mu = \bar{\psi}\Gamma_\mu\psi \tag{2.60}$$

where $\Gamma_\mu = \gamma_\mu$, $\sigma_{\mu\mu'}$, $\rho_{\mu'}$, ... Since $\psi = \psi_c$ (the phase is unimportant here) we have

$$\begin{aligned} O_\mu &= \frac{1}{2}\left[\bar{\psi}\Gamma_\mu\psi + \bar{\psi}_c\Gamma_\mu\psi_c\right] \\ &= \frac{1}{2}\left[\bar{\psi}\Gamma_\mu\psi + \psi^T\gamma_0{}^*C^+\gamma_0\Gamma C\bar{\psi}^T\right] \\ &= \frac{1}{2}\left[\bar{\psi}\Gamma_\mu\psi - \bar{\psi}C^T\Gamma^T\gamma_0^T C^*\gamma_0{}^+\psi\right] \\ &= \frac{1}{2}\left[\bar{\psi}\Gamma_\mu\psi + \bar{\psi}(C\Gamma_\mu^T C^+)\psi\right] \end{aligned} \tag{2.61}$$

Now

$$\begin{aligned} C\gamma_\mu^T C^+ &= -\gamma_\mu \\ C(\gamma_\mu\gamma_5)^T C^+ &= C\gamma_5^T\gamma_\mu^T C^+ = -\gamma_5\gamma_\mu = \gamma_\mu\gamma_5 \\ C(\sigma_{\mu\nu})^T C^+ &= \sigma_{\nu\mu} = -\sigma_{\mu\nu} \end{aligned} \tag{2.62}$$

Thus, the most general form for O_μ is

$$O_\mu = \bar{\psi}(a\gamma_\mu\gamma_5 + bp_\mu\gamma_5 + cp_\mu) \tag{2.63}$$

We can determine from (2.63) the electromagnetic current, using current conservation and taking ψ , $\overline{\psi}$ to be free spinors, satisfying the D.E.

$$0 = p_\mu \mathcal{O}_\mu = \bar{\Psi}\left[(2am + bp^2)\gamma_5 + cp^2\right]\psi \qquad (2.64)$$

thus $C = 0$, $a = - b \cdot p^2/2m$:

$$\mathcal{J}_\mu^{el} = \bar{\Psi}(p^2\gamma_\mu - 2mp_\mu)\gamma_5 F(p^2)\Psi . \qquad (2.65)$$

This shows that the electromagnetic coupling of Majorana particles is purely axial.

3. THEORETICAL IDEAS

In the standard model[4], left-handed and right-handed particles are treated completely independently. One is therefore free not to introduce a right-handed neutrino ν_R and no neutrino mass arises, if one only allows Dirac mass terms. If one introduces ν_R, then one is left with the difficulty of how to "understand" the small mass.

As we have seen, it is possible to introduce Majorana masses even if ν_R is absent. Since the left-handed ν_L implies ν_{RC}, a mass term $\nu_{RC}\,\nu_L$ is possible, but it violates fermion number. Since ν_L is together with e_L in an SU(2) doublet, we must form an SU(2) invariant which contains the Majorana mass term. It is not possible to write the Lorentz scalar

$$\mathcal{L}_M = (\nu_L^T e_L^T)\, C^{-1} \begin{pmatrix} \nu_L \\ e_L \end{pmatrix} * \text{Cl. Gord.} \tag{3.1}$$

because it vanishes by the Cl. Gordan coeff. of SU(2) which are ϵ_{ij}. Furthermore, it is not U(1) invariant (electric charge).

An alternative is to consider the "Majorana" Yukawa coupling *

$$\begin{aligned}\mathcal{L}_Y &= \text{const.}\, (\nu_L^T e_L^T)\, C^{-1} \begin{pmatrix} \nu_L \\ e_L \end{pmatrix} \phi \\ &= \text{const}\, (\nu_L C^{-1} \nu_L \phi_1 + 2\nu_L^T C^{-1} e_L \phi_0 + e_L^T C^{-1} e_L \phi_{-1})\end{aligned} \tag{3.2}$$

where ϕ_i is a scalar, and a SU(2) triplet; it carries hypercharge $Y = -1$ ((ν_L, e_L) has $\frac{1}{2}$).

We see that if $\langle\phi_1\rangle \neq 0$, then ν_L acquires a Majorana mass. Since (ν_L, e_L) are in a doublet, the fermion number, violated by the Majorana mass term is lepton number (more precisely: "electronic" lepton number). The field ϕ_1 carries (- two) units of it. Therefore, when $\langle\phi_1\rangle \neq 0$, a (massless) goldstone boson exists, since $\mathcal{L}_Y$ is invariant under $(\nu_L, e_L) \to e^{i\alpha} (\nu_L, e_L)$, $\phi \to e^{-2i\alpha} (\nu_L, e_L)$.

---

* One can assign ϕ zero or non-zero lepton number (in the latter case ϕ has lepton number-2, because (3.2) is invariant under

This is known as the "Majoron" (see Footnote and references therein). We note, that ϕ contains also a doubly charged field, ϕ_1. $<\phi_1>$ cannot be too large, in order not to upset the successful relation $m_W/m_Z = \cos\theta_W$.[17]

The situation in SU(5) [18] is similar. Due to its particular group structure, the representation which contain all necessary fields do not contain ν_R. Although this gives some rationale for the absence of ν_R, one can clearly introduce an SU(5) singlet field ν_R. Also in this theory, Majorana masses are the only possibility, and terms such as (3.2) can be constructed. Such terms were introduced for various reasons, such as in an attempt to understand the mass pattern of the observed fermions as a radiative succession[19]. The neutrino masses obtained are, however, very small.

Before continuing, we note that in both models, ν_R, being a gauge singlet, has no interactions other than through scalars. Furthermore, it is always possible to write a term $m \cdot \nu_R^T C^{-1} \nu_R$ (it is gauge invariant). Since m is a group singlet, one usually invokes the "survival" hypothesis[20] which says that m should be as large as M_P (Planck mass = 10^{19} GeV) because it does not vanish in any symmetry-limit (except fermion number). This will be important in the following.

---

* $(\nu_L, e_L) \to e^{i\alpha}(\nu_L, e_L)$; $\phi \to e^{-2i\alpha}\phi)$.

In the first case, lepton number is explicitely violated (see e.g. R. Barbieri, D.V. Nanopoulos, G. Morchio and F. Strocchi, Phys. Lett. 90B (1980) 91; M. Magg and C. Wetterich, Phys. Lett. 94B (1980) 61; Riazzudin, R.E. Marshak and R.N. Mohapatra, Phys. Rev. D24 (1981) 1310; T.P. Cheng and L.F. Li, Phys. Rev. D22 (1980) 2861).

In the second case, lepton number is spontaneously broken. See Y. Chikashige, R.N. Mohapatra and R.D. Peccei, Phys. Lett. 98B (1981) 265; G.B. Gelmini and M. Roncadelli, Phys. Lett. 99B (1981) 411. Such models lead to Goldstone bosons, connected with the broken fermion number ("Majorons"). They are similar in many respects to the axions. For further investigations of such models see e.g. H.M. Georgi, S.L. Glashow and S. Nussinov, Nucl. Phys. B297 (1981) 297. J. Schechter and J. Valle, Phys. Rev. D25 (1982) 774.

The situation changes if one considers models with a left-right symmetric structure which necessarily contain ν_R (a similar situation might exist in composite models). The low-energy model is a $SU(2)_L \times SU(2)_R \times U(1)$ based[21] theory, the grand unified version is the SO(10)[22,23] model. Both contain all usual fields and ν_R, which forms, with e_R, a doublet under $SU(2)_R$.*

We can describe the basic mechanism without much group theoretical detail. For one generation, the most general mass term (2.43) is of the form

$$\mathcal{L}_M = (\nu_L^T \nu_{LC}^T) C^{-1} \begin{pmatrix} a & b \\ b & c \end{pmatrix} \begin{pmatrix} \nu_L \\ \nu_{LC} \end{pmatrix} + h.c. \tag{3.3}$$

where "a" transforms as (3,1) under $SU(2)_L \times SU(2)_R$ (e.g. it is a triplet under $SU(2)_L$ and a singlet under $SU(2)_R$), b as (2,2) and c as (1,3). Of course, a, b, c are given by coupling constants times a v.e.v. of a scalar Φ with above group properties. In SO(10), all fermions, including ν_R, are in a 16-dimensional representation ψ_{16} (all left-handed, as described in (2.8)).[24] The Yukawa coupling giving rise to (3.3) is of the form $\psi_{16} C^{-1} \psi_{16} \phi$, thus ϕ must belong to $16 \otimes 16 = 10 \oplus 120 \oplus 126$. The 10 contains the (2,2) and thus b, the 126 the (3,1) and (1,3), that is a, c.

In its usual form, SO(10) has a ϕ_{10} and a ϕ_{16}. (The ϕ_{16} transforms like the ψ_{16}, in particular it has a component which transforms like ν_R, ϕ_{16}^1). ϕ_{16} breaks SO(10) to SU(5) with $\langle\phi_{16}\rangle \neq 0$, ϕ_{10} breaks $SU(2)_L \times U(1)$ and gives (Dirac) masses to fermions. Since all fermions are in one irreducible representation, one obtains $b = m_\mu$, and thus $m_\nu = m_\mu$, with Dirac neutrinos, $\nu = \nu_L + \nu_R$, since it appears that $a = c = 0$.

The above conclusion can be avoided if $c \neq 0$. Then, the eigenvalues are (2.44)

---

* Grand unification models are in a way a natural context for discussion of (Majorana) neutrino masses, since they tend to violate global conservation charges (Baryon number etc.).

$$\lambda_{1,2} = \frac{c}{2}\left\{1 \pm \sqrt{1 + \frac{4b^2}{c^2}}\right\} \simeq \begin{cases} -b^2/c & b \ll c \\ c & \end{cases} \tag{3.4}$$

and the mass eigenstates are (2.46) for $b \ll c$

$$\begin{aligned} \chi_1 &\simeq (\nu_L + \nu_{RC}) - \frac{b}{c}(\nu_R + \nu_{LC}) \\ \chi_2 &\simeq (\nu_R + \nu_{LC}) + \frac{b}{c}(\nu_L + \nu_{RC}) \end{aligned} \tag{3.5}$$

We can now discuss the magnitudes of a, b, c once we have recognized the mechanism.

Originally, a ϕ_{126} was introduced, to provide a term c .[23] Since the corresponding v.e.v. breaks SO(10) to SU(5) , the v.e.v. is very large, about 10^{16} GeV . Then, c =(Yukawa coupling) • 10^{16}GeV. Using $b = m_\mu \cong 5$ MeV , $b^2/c \cong 10^{-8}$ eV for a small Yukawa coupling of 10^{-4} . We emphasize that the Yukawa coupling is unknown; we do not even know whether it is larger or smaller for other generations. In view of the small masses, we do not expect observable effect, even though the mixing angles can be large. In the usual estimates one takes a "typical" b of ≅ 1 GeV and obtain $b^2/c \cong 10^{-3}$ eV.

Subsequently it was observed that ϕ_{126} is not needed,[25] since the term c can also be an "effective" 126 arising from a higher order effect. Indeed, since $16 \otimes 16 = 126 \oplus \ldots$, we can use ϕ_{16} to obtain an effective term $(\psi_{16}C^{-1}\psi_{16})(\phi_{16}\phi_{16})\frac{1}{M}$, where M is a typical scale.

It is straightforward to convince oneself that above term only arises at the two loop level. Then, we expect $c \cong (\alpha^2)\,10^{16}$ where α is a typical "fine structure constant" at 10^{16} GeV , $\alpha \cong 2\cdot10^{-2}$. With this $\frac{b^2}{c}$ few eV, a range which is accessible to experiments if mixings occur (oscillations).

We note that, in general, c corresponds to the mass of W_R , b is related to m_{W_L} . Thus, in $SU(2)_L \times SU(2)_R$ - type theories, $\frac{b^2}{c} \cong m_q^2/m_{W_R}$. In $SU(2)_L \times U(1)$ or SU(5) type theories, $b^2/c \cong \frac{m_q^2}{M_P}$ as discussed previously, and one expects the resulting neutrino mass to be very small.

One can also view the relation $b^2/c \cong m_q^2/c$ as a lower bound on c, using the upper bounds on m_ν.[26] This gives possible ranges for existence of heavy neutrinos. For instance[26]

$$m(\nu_{R\,\text{electron}}) \gtrsim 4\,\text{GeV}\ ^{*}. \tag{3.6}$$

In SU(2) x U(1), the heavy neutrino would, however, decouple (except for Higgs interactions) and is hard to see.

In alternative to the above mechanism is the following.[27] We can extend the particle contents of $SU(2)_L \times SU(2)_R \times U(1)$ or SO(10) by adding a field, S_L which is a singlet under the gauge group (and left-handed). If it is defined to have fermion number +1, then the most general Dirac mass term is (2.43)

$$\mathcal{L}_M = (\bar{\nu}_{RC}\ \bar{S}_{RC}\ \bar{\nu}_R)\begin{pmatrix} 0 & 0 & a \\ 0 & 0 & b \\ a & b & 0 \end{pmatrix}\begin{pmatrix} \nu_L \\ S_L \\ \nu_{LC} \end{pmatrix} + h.c. \tag{3.7}$$

where $a \sim (2,1)$ and $b \sim (1,2)$ under $SU(2)_L \times SU(2)_R$. Thus, we expect $a/b \simeq m_W/m_{WR}$. (3.7) leads clearly to a massless and a massive Dirac state:

$$m = 0 \qquad \nu_1 = \cos\alpha\,\nu_L - \sin\alpha\,S_L \tag{3.8.1}$$

$$m = \sqrt{a^2+b^2} \qquad \nu_2 = \sin\alpha\,\nu_L + \cos\alpha\,S_L + \nu_R \tag{3.8.2}$$

$$\operatorname{tg}\alpha = \frac{a}{b}$$

---

* From a theoretical view-point these numbers are not very appealing, because one tends to associate $m(\nu_R)$ with a scale of new physics - probably far away from 4 GeV. However, a very tiny Yukawa coupling could well give (3.6). From low energy experiments (such as double β-decay, $K^+ \to \pi^-\mu^+\mu^+$ (see below ref. 76)) it is impossible to rule out a Majorana neutrino in the few GeV-range.

The only direct phenomenological consequence of this scheme is a violation of Cabibbo-KM-universality[4], since only ν_1 participates in ordinary weak processes (ν_2 is too heavy, its mass being coupling x M_{W_R}), and therefore the coupling $We\nu$ is $(\cos\alpha \cdot g)$ instead of g. However, already for $\alpha \leqq {}^1/_{10}$, this effect is unmeasurable. This scheme allows a massless neutrino without a very large mass scale, since (3.8.1) is true for all a, b. The state ν_2 could be in an interesting mass range (10-100 GeV) if $m_{W_R} \cong 10$ TeV and the Yukawa coupling is 10^{-3}-10^{-2} as for usual fermions. Its production, in present $p\bar{p}$-colliders, is however suppressed, by a factor $\sin^2\alpha$, ($<10^{-2}$) since S_L is a gauge singlet and ν_R couples only via W_R. Only a very light scalar (Higgs) with enhanced coupling might give sizeable production.

This scheme (as well as the previous) gives rise to "oscillatory" effects if there are many generations. We will discuss this later.

The two schemes above gave either a light Majorana particle or a massless particle. We can obtain a light Dirac particle by combining the two ideas in the context of the $SU(2)_L \times U(1)$ model.[28] Since we want a light and a heavy Dirac particle, we need four states. Consequently, (3.7) is extended to include S_R, a gauge group singlet. We set

$$\mathcal{L}_M = \left(\bar{\nu}_{RC}\ \bar{S}_{RC}\ \bar{\nu}_R\ \bar{S}_R\right)\begin{pmatrix} 0 & 0 & 0 & m_1 \\ 0 & 0 & m_2 & M \\ 0 & m_2 & 0 & 0 \\ m_1 & M & 0 & 0 \end{pmatrix}\begin{pmatrix} \nu_L \\ S_L \\ \nu_{LC} \\ S_{LC} \end{pmatrix} + h.c. \tag{3.9}$$

$m_1 \sim (2,1)$ m_2, $M \sim (1,1)$ under $SU(2)_L$. In order that the entry $(\bar{\nu}_R \nu_L)$ vanished, a new, global symmetry had to be imposed, under which ν_R transforms non-trivially; then m_2 also transforms non-trivially under this symmetry.

(3.9) gives rise to two Dirac particles; if m_1, $m_2 \ll M$ then

$$\psi_1 \simeq \left(\nu_L - \frac{m_1}{M} s_L\right) + \left(\nu_R - \frac{m_2}{M} s_R\right) \quad m_1 \simeq \frac{m_1 m_2}{M} \tag{3.10.1}$$

$$\psi_2 \simeq \left(s_L + \frac{m_1}{M} \nu_L\right) + \left(s_R + \frac{m_2}{M} \nu_R\right) \quad m_2 \simeq M \tag{3.10.2}$$

Again we must discuss the magnitudes of m_1, m_2, M . Apart from Yukawa couplings, $m_1 \sim O(M_W)$, $M < M_P$ (see ref. 20), whereas m_2 is associated with the breaking scale of the global symmetry. On this we can only speculate. If the global symmetry is part of a Peccei-Quinn[29] symmetry, then $m_2 \quad 10^{10}$ GeV . This is because, the Goldstone boson associated with the breaking of the Peccei-Quinn symmetry, the axion, can carry energy which might affect adversly the energy balance of the universe unless it couples enough to matter; the coupling being proportional to $^1/m_2$.[10] Using these numbers, $m(\psi_1) < 10^2 \cdot 10^{10} \cdot 10^{-19}$ GeV $< 10^2$ eV , a number preferred by cosmological considerations and the ITEP-experiment.* We will discuss the many-generation phenomenology later.

We will next turn to experimental manifestations of neutrino masses.

---

* See later for references

4. OSCILLATIONS[31]

If the neutrinos are massive, then it is possible, that, like the quarks, "weak interaction eigenstates" (e.g. those states produced in a weak process; such as β -decay) are not eigenstates of the Hamiltonian (or mass eigenstates), the physical states, but linear combinations thereof:

$$|\psi(0)\rangle \equiv |weak_e\rangle = \sum_i C_{ei}\, |phys_i\rangle \tag{4.1}$$

Assume now that the state $|weak_l\rangle$ is produced at $t = 0$, and moves freely. Every component in (4.1) develops as

$$|phys_i\rangle = e^{-iHt}\,|phys_i\rangle = e^{-iE_it}\,|phys_i\rangle \tag{4.2}$$

so, for $t \neq 0$, (4.1) becomes

$$|\psi(t)\rangle = \sum_i C_{ei}\, e^{-iE_it}\,|phys_i\rangle \tag{4.3}$$

which in general is different from (4.1), with an oscillatory pattern.[31] By investigating $|\psi(t)\rangle$ (that is measuring its components $|weak_l\rangle$ through a weak interaction process) we obtain information on the C_{li} and E_i .

The general expression that connects $|weak_l\rangle$ and $|phys_i\rangle$ is (2.52) (χ are the (Majorana) mass eigenstates)[1]:

$$|phys_i\rangle \sim \chi = U\begin{pmatrix}\psi_L\\ \psi_{LC}\end{pmatrix} + SU^*\begin{pmatrix}\psi_{RC}\\ \psi_R\end{pmatrix} \tag{4.4}$$

where ψ_L, ψ_R are the weak eigenstates and ψ_{RC}, ψ_{LC} are their complex conjugates. U is a $(n_L+n_R)^2$ unitary matrix:

$$U = \begin{pmatrix}A & B\\ C & D\end{pmatrix} \qquad \begin{array}{l} A : n_L \times n_L \\ D : n_R \times n_R \\ B, C^T : n_L \times n_R \end{array} \tag{4.5.1}$$

and we define

$$\chi_{(L)} = (A\psi_L + B\psi_{LC}) + SA^*\psi_{RC} + SB^*\psi_R \qquad (4.5.2)$$
$$\chi_{(R)} = (C\psi_L + D\psi_{LC}) + SC^*\psi_{RC} + SD^*\psi_R$$

S is a diagonal matrix of phases; e.g. in (4.5.2) S refers to the phase associated with the corresponding χ. Special cases are (compare with (2.43))

a) $$M_D = 0 \qquad B = C = 0 \qquad (4.6)$$

$$\chi_{(L)} = A\psi_L + SA^*\psi_{RC} \qquad (4.7)$$
$$\chi_{(R)} = D\psi_R + SD^*\psi_{LC}$$

b) $$M_L = M_R = 0 \qquad (4.8)$$

$$\mathcal{U} = \begin{pmatrix} A & -B \\ A & B \end{pmatrix} \qquad (4.9)$$

with $A^T M_D B = \text{diag}$. If $\chi_\pm \equiv \chi_{(L),(R)}$ then

$$\chi_- = (A\psi_L - B\psi_{LC}) + S_-(A^*\psi_{RC} - B^*\psi_R) \qquad (4.10)$$
$$\chi_+ = (A\psi_L + B\psi_{LC}) + S_+(A^*\psi_{RC} + B^*\psi_R)$$

where χ_+, χ_- have equal masses. $S_\pm = \pm 1$ in this case, as we know (2.51). If $n_L \neq n_R$, say $n_L < n_R$ then there are $(n_R - n_L)$ massless modes (2.59), given by the last (say) $(n_R - n_L)$ rows of B . Only the first n_L rows of B must then be used in (4.10): $\widehat{B}$. Instead of (4.10) one can just write

$$\psi = A\psi_L + \widehat{B}\psi_R \qquad (4.11)$$

and consider them as ordinary Dirac spinors.

The weak interaction states ν_1 are defined by their interaction:

$$\mathcal{L} = g_L J^+_{\mu L} W_L^{\mu} + g_R J^+_{\mu R} W_R^{\mu} + g_{Ni} J^0_{\mu i} Z_i^{\mu} + h.c. \tag{4.12}$$

where $J^+_{\mu L,R}$ are the left- and right-handed charged and neutral currents, given by $(l_1 = e, l_2 = \mu, \ldots)$

$$J^+_{\mu L,R} = \sum_{\ell} \bar{e}_{\ell} \gamma_{\mu} \binom{L}{R} \nu_{\ell} \quad J^0_{\mu i} = \sum_{\ell} \bar{\nu}_{\ell} \gamma_{\mu} (a_i L + b_i R) \nu_{\ell} \tag{4.13}$$

and $W^{\mu}_{L,R}$, $Z^{\mu}_{L,R}$ are the respective intermediate bosons. In writing (4.13), we have assumed that Higgs-interactions are very suppressed. In the following we will only take W_L and one Z, since W_R, Z are presumably much heavier than W_L, Z. We therefore must solve (4.4) for $\nu_L \equiv \psi_L$:

$$\nu_L = \{A^+ \chi_{(L)} + C^+ \chi_{(R)}\}_L \tag{4.14}$$

The Dirac equation for $\chi_{(L),(R)}$ is

$$\not{p}\, \chi_{\substack{(L)\\(R)}}\Big|_L = m\, \chi_{\substack{(L)\\(R)}}\Big|_R \tag{4.15}$$

which we will use later.

Using (4.14) we get

$$J^+_{\mu L} = \sum_{\ell} \bar{e}_{\ell} \gamma_{\mu} (A^+ \chi_{(L)} + C^+ \chi_{(R)})_L \tag{4.16}$$

$$J^0_{\mu L} = \sum (\bar{\chi}_{(L)} A + \bar{\chi}_{(R)} C)_L \gamma_{\mu} (A^+ \chi_{(L)} + C^+ \chi_{(R)})_L \tag{4.17}$$

(4.17) is noteworthy. Since A, C are generally not unitary, $J_{\mu L}$ is not flavor-diagonal, although we may have started with a theory where the ψ_L have the same transformation-property under the gauge group, and also the ψ_R. This seemingly violates a general result.[32] The point, however, is that this result holds only if the masses are only Dirac or (purely) Majorana. If both Majorana and

Dirac masses are present, then there are effectively two left-handed fields, ψ_L and ψ_{LC} which can mix, and then, since the currents for ψ_L and ψ_{LC} are generally different, $J_{\mu L}$ is flavor non-diagonal.

This is easily checked: if $M_D = 0$, A = unitary, C = 0 ; if $M_L = M_R = 0$, A = C = unitary. Since $\chi_\pm = \chi_{(L),(R)}$ have the same mass, we can use $(\chi_{(L)} + \chi_{(R)}) = \psi_S$ as new field and $J_{\mu L} = 2\overline{\psi}_{SL} A\gamma_\mu A^+\psi_{SL} = 2\overline{\psi}_{SL}\gamma_\mu\psi_S$. Equally, in the charged current, we can use ψ_S .

We now proceed to derive the oscillation formulae. At t = 0 the state (we omit index 1)

$$|\psi(0)\rangle \equiv \nu_L = (A^+\chi_{(L)} + C^+\chi_{(R)})_L \tag{4.18.1}$$

$$+ \frac{1}{E_i}\left[A^+ m_{(L)}\chi_{(L)} + C^+ m_{(R)}\chi_{(R)}\right]_R \tag{4.18.2}$$

is produced. The second term comes from the Dirac equation, Eq (4.15). Now, at t , each of the states $\chi_{(L),(R)\ldots}$ develops according to (4.2). Using (4.4) to re-express the χ in terms of the ψ (since the detection, being a weak interaction process only "sees" the ψ) we have[33]

$$\begin{aligned}|\psi(t)\rangle = {} & A^+e^{-iEt}(A\psi_L + B\psi_{LC}) + C^+e^{-iEt} \times \\ & \times (C\psi_L + D\psi_{LC}) \\ & + \frac{1}{E}\Big\{A^+m_{(L)}e^{-iEt}(SA^*\psi_{RC} + SB^*\psi_R) \\ & + C^+m_{(R)}e^{-iEt}(SC^*\psi_{RC} + SD^*\psi_R)\Big\} =\end{aligned} \tag{4.19}$$

$$
\begin{aligned}
&= (A^+ e^{-iE_{(L)}t} A + C^+ e^{-iE_{(R)}t} C)\,\psi_L \\
&+ (A^+ e^{-iE_{(R)}t} B + C^+ e^{-iE_{(R)}t} D)\,\psi_{LC} \\
&+ \frac{1}{E}(A^+ m_{(L)} S e^{-iE_{(L)}} A^* + C^+ m_{(R)} e^{-iE_{(L)}t} SC^*)\,\psi_{RC} \\
&+ \frac{1}{E}(A^+ m_{(L)} S e^{-iE_{(L)}} B^* + C^+ m_{(R)} e^{-iE_{(R)}t} SD^*)\,\psi_R
\end{aligned}
\tag{4.20}
$$

with $E_{(L),(R)} = m^2_{(L),(R)} + \vec{p}^2$, $\vec{p}^2$ fixed.
Considering now the detection of the $\psi_{L,LC,RC,R}$ by charged currents, we observe that ψ_L will give rise to a lepton produced in the detection, ψ_{RC} to an antilepton; ψ_{LC} , ψ_R are only visible, if there are right-handed currents. Neutral currents are also possible, but not always of interest, because they tend not to distinguish different neutrinos.

(4.20) gives rise to a number of selection rules. If $M_D = 0$, $B = C = 0$, and only ψ_L or ψ_{RC} can be produced. If $M_L = M_R = 0$, then, since $E_{(L)} = E_{(R)}$, $C = A$, $D = -B$, ψ_{LC} cannot be produced; equally ψ_{RC} , because $S\chi_{(L)} = -S\chi_{(R)}$. Transitions from particle to antiparticle are called second class* oscillations, and occur, for $\frac{m}{E} \ll 1$, only if both M_D and M_M do not vanish. Particle-particle oscillations are called first class.

Using (4.20) we can calculate the oscillation probabilities. We have, using the form (4.5)

$$
\begin{aligned}
P(\ell,\ell',t) &= P((\nu_L)_\ell, (\nu_L)_{\ell'}, t) = \\
&= |A^+_{\ell k} A_{k\ell'} e^{-iE_{(L)k}t} + C^+_{\ell k} C_{k\ell'} e^{-iE_{(R)k}t}|^2 \\
&\equiv U^+_{\ell k} U_{k\ell'} e^{-iE_k t}\, U_{k'\ell} U^+_{\ell' k'} e^{-iE_{k'} t} \\
&\quad (U^+_{\ell k} \equiv A^+_{\ell k}\ (k=1,\dots n_L)\ ,\ C^+_{\ell k}\ (k=n_L+1,\dots,n_L+n_R))
\end{aligned}
\tag{4.21}
$$

* These were the oscillations originally considered by Pontecorvo(ref.31)

Since $(U_{k'l}\, U_{l'k'})^* = U_{k'l'}\, U^+_{lk'}$ we have

$$P(l,l',t) = \sum_k C^k_{ll'}\, C^{k'*}_{ll'}\, e^{i(E_{k'}-E_k)t}, \quad C^k_{ll'} = U^+_{lk} U_{kl'} \qquad (4.22)$$

Similarly, one calculates for instance

$$P(l,\tilde{l}',t) = P((\nu_L)_l, (\nu_{RC})_{l'}, t) =$$

$$\frac{U^+_{lk} U^*_{kl'} U_{k'l} U_{k'l'} S_{kk} S^*_{k'k'} e^{-it(E_k-E_{k'})} m_k m_{k'}}{E_k E_{k'}} \qquad (4.23)$$

etc.
We notice, that the $(L \to L)$ transitions are independent of the phases S. This can be seen to be a general result. From (2.51) we know that

$$\sqrt{S}\,U = \sqrt{S'}\,U' \qquad (4.24)$$

if $U \to U'$, or

$$U' = PU \qquad (4.25)$$

where P is a diagonal phase matrix. Under (4.25) the $C^k_{ll'}$ are unchanged. Under (4.25), the expression (4.23) changes by $P^2_{k'k'} P^{*2}_{kk}$. We can express this differently:

A $n \times n$ unitary matrix U has n^2 parameters, $\frac{n(n-1)}{2}$ of which are real angles, and $\frac{n(n+1)}{2}$ which are phases. In the current J^+, a unitary matrix U appears as

$$J^+_{\mu L} = \bar{a}_i U_{ij} \gamma_\mu L b_j$$

where a_i, b_i are differently charged physical fields. If they are both of Dirac-type, the transformation $b \to e^{i\beta}b$, $a \to e^{i\alpha}a$ does not change the mass terms ($m\bar{b}b$ etc.) and $2n-1$ (overall phase) phases can be removed.[34] If, say b_j, is Majorana,

$b \to e^{i\beta} b_j$ is not allowed, since it changes the mass, or the phase S which is observable $(U \to PU)$. Therefore, for the current (4.16), only n phases can be absorbed; U has therefore $\frac{(n+1)n}{2} - n = \frac{n(n-1)}{2}$ phases. The preceeding argument shows that (4.22) is also independent under the extra Majorana phases, but not (4.23).[35] From (4.22), (4.23) and similar expressions, we can derive relations between probabilities. (4.22) implies for real U :

$$P(\ell,\ell',t) = P(\ell',\ell,t) \quad \text{etc.} \tag{4.26}$$

The preceeding analysis assumes that the various neutrinos are stable. Obvious modifications are necessary for decaying neutrinos. Also, we note that if a certain state χ is too heavy to be produced by the initial ν (which has a certain energy) it does not occur in the sums (4.22) or (4.23). We will look briefly at this possibility later.

In the following we restrict ourselves to (4.22) since most experiments are analysed in this way, usually with two states. (4.22) now implies

$$\begin{aligned} P(\ell,\ell',t) = \sum_k |C^k_{\ell\ell'}|^2 + 2\,\mathrm{Re}\sum_{k,k'} C^k_{\ell\ell'} C^{k'*}_{\ell\ell'} \cos(E_k - E_{k'})t \\ \mp 2\,\mathrm{Im}\sum_{k,k'} C^k_{\ell\ell'} C^{k'*}_{\ell\ell'} \sin(E_k - E_{k'})t \,. \end{aligned} \tag{4.27}$$

For $n = 2$, there is no observable phase (above discussion): $(s = \sin\theta, \ldots)$

$$U = \begin{pmatrix} c & -s \\ s & c \end{pmatrix} \quad \begin{matrix} C^1_{11} = c^2 & C^1_{12} = -cs & C^2_{22} = s^2 \\ C^2_{11} = s^2 & C^2_{12} = -cs & C^2_{22} = c^2 \end{matrix} \tag{4.28}$$

$$\begin{aligned} P(1,2,t) &= 2c^2 s^2 \left(1 - \cos t\,(E_1 - E_2)\right) = \\ &= \sin^2 2\theta \,\sin^2\left(\frac{E_1 - E_2}{2} t\right) \end{aligned} \tag{4.29}$$

$$P(1,1,t) = 1 - P(1,2,t)$$

In all typical situations $E_i >> m_i$. We therefore have $E_1 - E_2 = \sqrt{p^2+m_1{}^2} - \sqrt{p^2+m_2{}^2} \cong \frac{m_1{}^2-m_2{}^2}{2p}$. Since in time t the neutrinos travel a distance $L = tv$; $v = \frac{p}{E}$, we get

$$P(1,1,t) = 1 - \sin^2 2\theta \sin^2\left(\frac{\Delta m^2 L}{E} 1.27\right)$$
$$P(1,2,t) = \sin^2 2\theta \sin^2\left(\frac{\Delta m^2 L}{E} 1.27\right) \tag{4.30}$$

where $\Delta m^2 = m_1{}^2 - m_2{}^2$ in eV

L = distance travelled in meters

E = initial energy in MeV

These units have been chosen because they correspond to most experimental values.

We note, however, that if m is not so small, we get $v = \frac{|p|}{E} \cong 1 - \frac{m^2}{2p^2}$, and at a distance L two neutrinos of masses m_1 , m_2 would arrive with time difference $L/V_1 - L/V_2 \cong (m_1{}^2-m_2{}^2)L/p^2$, which has been considered in supernova explosions.[36]

Before continuing we comment on the possibility that some neutrinos are too heavy to be produced. (4.18.1) and (4.4) give

$$|\psi(0)\rangle = (\nu_L)_e = U^+_{ei}\chi_{iL} = U^+_{ei}U_{ie'}(\nu_L)_{e'} \tag{4.31}$$

This gives seemingly $\delta_{11'}$ and thus $|\psi_0\rangle$ is indeed $(\nu_L)_1$. This does not hold, if some of χ_i are too heavy to be included in the sum. Then we must replace (4.31) by

$$|\psi(0)\rangle = \sum_{\substack{\text{only}\\\text{light}\\\text{states}}} U^+_{ei}U_{ie'}(\nu_L)_{e'} \neq \delta_{ee'}(\nu_L)_e \tag{4.32}$$

and we see that already $|\psi_0\rangle$ contains admixtures of $(\nu_L)_{1'}$: we say that $(\nu_L)_1$ and $(\nu_L)_{1'}$ are not orthogonal. This gives rise to a constant flux of $(\nu_L)_{1'}$, independent of L or t . This situation occurs in models such as discussed in the previous chapter where there is always a very heavy neutrino. A discussion of this effect can be found for instance in ref. 15.

The experiments searching for neutrino oscillations consist of a neutrino source emitting a certain neutrino type with an energy spectrum $N(E,L=0)$ (flux of neutrinos with energy E). At a certain distance, a detector is placed in which the neutrinos can undergo a weak process. One prefers charged currents (e.g. production of a charged lepton), because neutral currents do not distinguish certain neutrinos (ν_{Le} and $\nu_{L\mu}$) and fewer variables can be measured. In this way, $N(E,L)$ is measured (typically L is fixed). If the spectrum of the source $N(E,L=0)$ is known, a comparison with (4.29), (4.30) is possible. Depending on the source, "appearance" experiment $(\nu_l \to \nu_l)$ or "disappearance" experiments $(\nu_l \to \nu_{l'})$ can be made.

(4.30) implies:

$\Delta m^2 \gg \frac{1.27L}{E}^{-1}$: rapid oscillations which are not resolvable (finite energy resolution); one should see a constant reduced flux of neutrinos ($\frac{1}{2}$ for two neutrinos, $\frac{1}{n}$ for n) apart from the factor $\sin^2 2\theta$.

$\Delta m^2 \ll \frac{1.27L}{E}^{-1}$: The factor $\sin^2(\frac{\Delta m^2 L}{E}\, 1.27)$ essentially vanishes; $P(\nu_l \to \nu_{l'}) \cong 0$, $(\nu_l \to \nu_l) \cong 1$.

$\Delta m^2 \cong \frac{1.27L}{E}^{-1}$: clearly visible oscillations.

These considerations clearly show that an experiment is most sensitive to small Δm^2, if E is small and L is large. The following table gives the typical figures for various neutrino sources.[37]

Source	Neutrino-type L=0	Energy	Length L	Flux at detector (ν/cm^2S)	$\frac{E}{1.27L}$ $(eV)^2$
Reactor	ν_{RCe}	1 MeV	10-60 m	10^{12}	10^{-2}
Sun	ν_{Le}	0.1 MeV	10^{11} m	10^{10}	10^{-12}
Accelerator	$\nu_{L\mu}$, $\nu_{RC\mu}$	1 GeV	10^4 m	10^4	10^{-1}
Meson factory	$\nu_{L\mu}$ $\nu_{RC\mu}$ ν_e	10 MeV	10 m	10^7	10^{-1}

Table 4.1: Some Neutrino sources and their properties

We see that in reactor or solar experiments, which are the most sensitive, only disappearance experiments are possible because the energy of the neutrino is not sufficient to produce a muon (or even τ) in the detector.

a) reactor experiments

The typical setup of a reactor experiment is shown in Fig. 4.1.

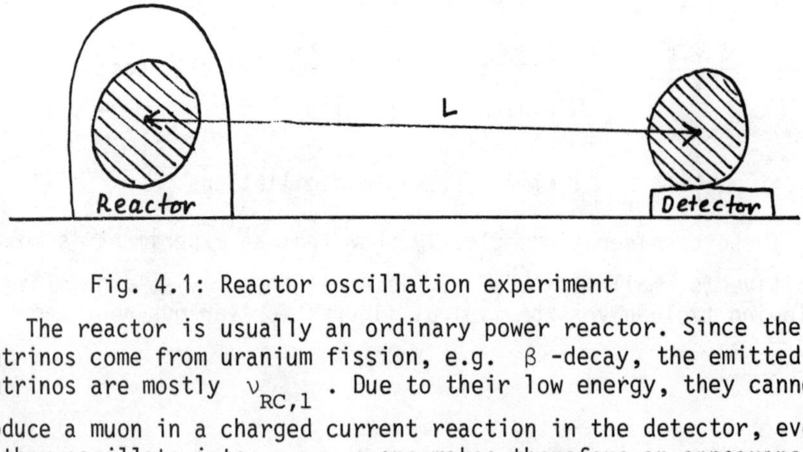

Fig. 4.1: Reactor oscillation experiment

The reactor is usually an ordinary power reactor. Since the neutrinos come from uranium fission, e.g. β -decay, the emitted neutrinos are mostly $\nu_{RC,1}$. Due to their low energy, they cannot produce a muon in a charged current reaction in the detector, even if they oscillate into $\nu_{RC,\mu}$; one makes therefore an appearance experiment. Table 4.2 shows reactions that have been used to detect the $\nu_{RC,1}$.[38]

Reaction	Cross-section	threshold
$\bar{\nu}+p \to n+e^+$	63.4	1.8 MeV
$\bar{\nu}+d \to n+n+e^+$	1.13	4.0 MeV
$\bar{\nu}+d \to n+p+\bar{\nu}$	3.10	2.3 MeV
$\bar{\nu}+e^- \to \bar{\nu}+e$	0.37	1.0 MeV

Table 4.2: Processes to detect $\bar{\nu}$. The cross sections are in 10^{-44} cm²/fission (for ^{235}U); the last reaction (neutral current) is for $\sin^2\theta_W = 0.25$ and averaged electron energies (from ref. 37).

The first reaction is the most favorable; it combines a low threshold and a large cross-section. However, the deuteron reactions have also been used. Because the neutral current process is insensitive to the neutrino species (since all $\bar{\nu}_{RC}$ have the same coupling to Z (however, if there are oscillations $\nu_L \to \nu_{LC}$ (or $\nu_{RC} \to \nu_R$) as discussed before (second class), then, because ν_R does not couple to Z being a $SU(2)_L \times U(1)$ singlet), this is not true), it can be used, to "calibrate" the total neutrino flux. Mostly, the first reaction in the table is used.

The yield $Y(L,E_e)$ of electrons of a certain energy as some distance L is clearly proportional to the number of neutrinos at L ; we write, using (4.30)[39]

$$Y(L,E_e) = Y_{no\,osc}(L,E_e)\left(1-\sin^2 2\theta \sin\left(\frac{\Delta m^2 L}{E_{\bar{\nu}}}1.27\right)\right) \tag{4.33}$$

where $Y_{no\ osc}$ is the yield in absence of oscillations, given by

$$Y_{no\,osc} = \frac{S(E_{\bar{\nu}},L=0)}{4\pi L^2}\,\sigma(E_{\bar{\nu}},E_e)\,N_{PT} \tag{4.34}$$

where $S(E_{\bar{\nu}},L=0)$ is the initial (at the reactor) spectrum of antineutrinos, $\frac{1}{4\pi L^2}$ the "geometrical thinning" factor, $\sigma(E_{\bar{\nu}},E_e)$ the weak cross-section in the detector (see table 4.2) and N_{PT} the number of protons in the target. We notice that if we know $Y_{no\ osc.}$ at some distance L', then $Y_{no\ osc}$ at L is different by a factor $(L'/L)^2$.

The reactor flux is difficult to calculate[40] and many complications arise. If one could measure $Y(L,E)$ at various distances, then, by forming $R(L,L')\ Y(L,E)/Y(L',E)$, $S(E_{\bar{\nu}},L=0)$ can be eliminated:

$$R(L,L') = \left(\frac{L'}{L}\right)^2 \frac{1-\sin^2 2\theta \sin^2\left(\frac{\Delta m^2 L}{E} 1.27\right)}{1-\sin^2 2\theta \sin^2\left(\frac{\Delta m^2 L}{E} 1.27\right)} . \qquad (4.35)$$

However, the detection system is very heavy, and cannot be moved easily, also, one must measure several months at one L to accumulate enough statistics. Therefore, only 2-3 distances, far apart, have been measured so far; it is clearly impossible to measure out (4.30) by varying L over a large number of values. From (4.35) we see that this method is insensitive to oscillations, if $\Delta m^2 >> \frac{E}{L,L'\ 1.27}$, or if $\frac{L'\Delta m^2}{E} 1.27 = \frac{L\ \Delta m^2}{E} 1.27 + 2\Pi$ etc.

One therefore attempts to measure (4.30) by calculating as good as possible $S(E_{\bar{\nu}}, L=0)$[40] and measuring the energy of the neutrinos by determining the positrons momentum and that of the recoiling neutron. The present data include results obtained in this way as well as results from measurements at 2(3) different distances.

The setup (Fig. 4.1) and the usual experimental difficulties complicate our above straight-forward analysis. Due to the finite extension of reactor and detection system, the length L is not unique; the energy acceptance and resolution must also be taken into considerations. One replaces (4.35) by[39]

$$\bar{Y}(\bar{E}_{\bar{\nu}},\bar{L}) = \int dL'\, dE'\, Y(L',E'_e)\, h(L,L')\, a(E'_e) \qquad (4.36)$$

where h, a describe the distance distribution and energy acceptance. Fig. 4.2 shows the effect of h on the oscillation pattern.[41] Similar effects come from finite energy resolution and acceptance.

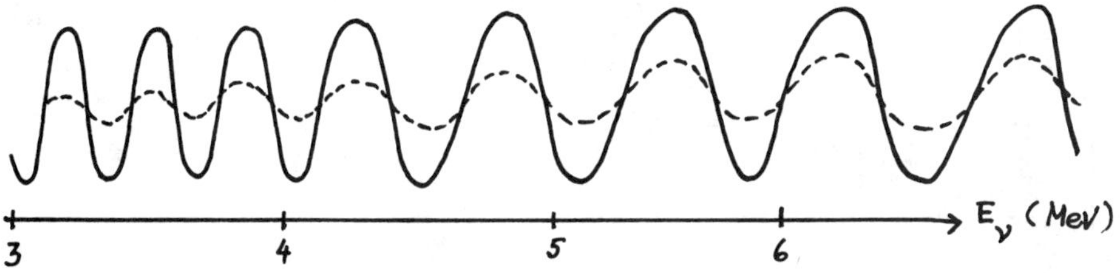

Fig. 4.2: "Washout of the oscillations due to various length, at the distance L = 37,9m , Δm^2 = 2 eV in the Gösgen experiment (from ref. 41)

Fig. 4.3 illustrates the testable Δm^2 (at sm $2\theta = 1$) at the various experiments and the effects of the finite energy resolution (from ref. 41)

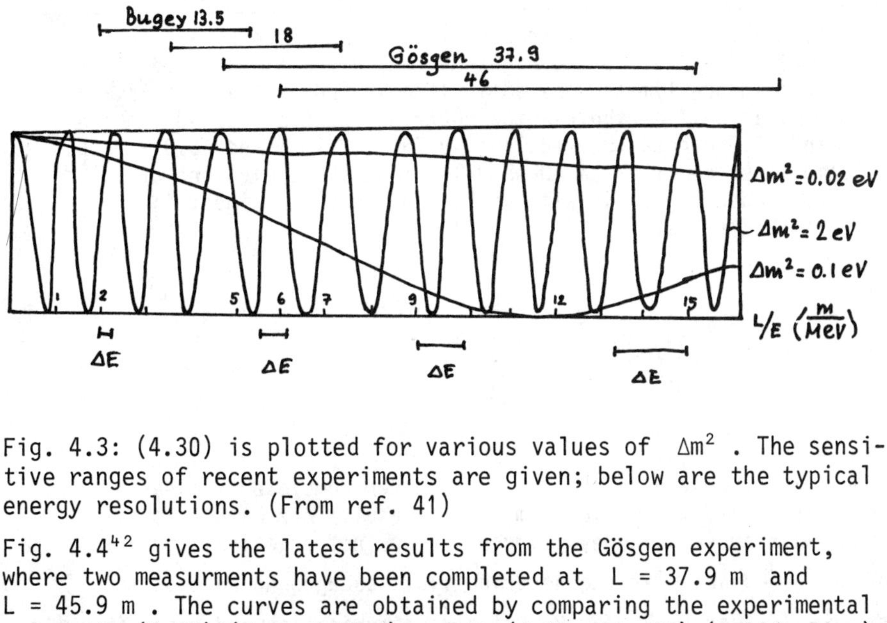

Fig. 4.3: (4.30) is plotted for various values of Δm^2 . The sensitive ranges of recent experiments are given; below are the typical energy resolutions. (From ref. 41)

Fig. 4.4[42] gives the latest results from the Gösgen experiment, where two measurments have been completed at L = 37.9 m and L = 45.9 m . The curves are obtained by comparing the experimental values to (4.35) (dashed line) and to (4.33, 34, 36) (solide line). The regions to the right of the curves are excluded. The strong oscillatory behaviour is explained in Fig. 4.5.

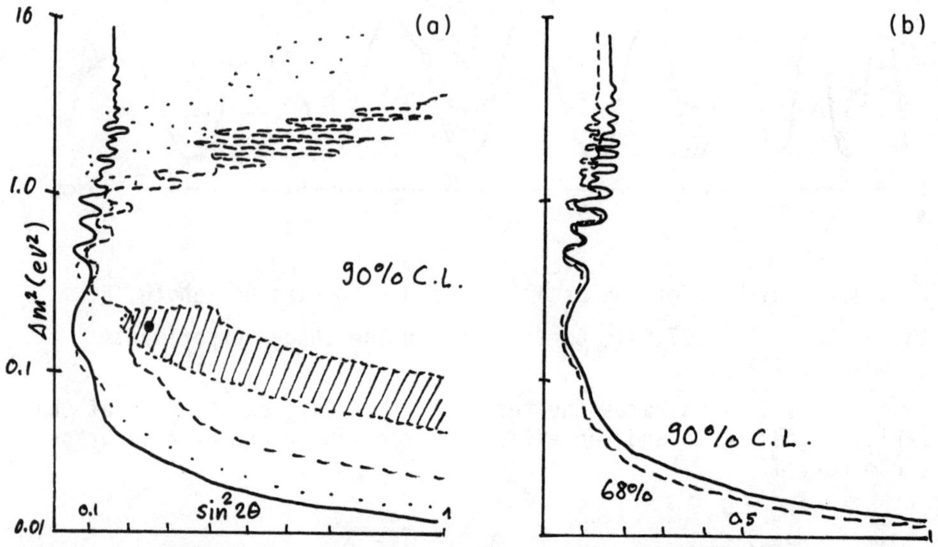

Fig. 4.4: Limits on the neutrino oscillation parameters Δm^2 versus $\sin^2 2\theta$. The area to the right of the curves is excluded. (a) 90% confidence limits. The solid line includes data from 37.9 and 45.9 m, as well as the reactor neutrino spectra. Dashed line: 37.9 m and 45.9 m data. Dotted line: 8.76 m, 37.9 m and 45.9 m data. (b) Solid line: see fig. 2a. Dashed line: 68% CL equivalent of solid line. The shaded region is allowed by the Bugey-experiment[43]; the dot indicate its best fit. (Fig. 4.6)*

---

* A new run at Gösgen at $L = 64$ m has been completed increasing the sensitivity to smaller Δm^2 . The results are hardly consistent with those in Bugey. Larger L are not practical; because of the $\frac{1}{L^2}$ factor in the neutrino flux, background becomes too strong. (G. Zacek, V. Zacek, Private communication and paper in preparation.)

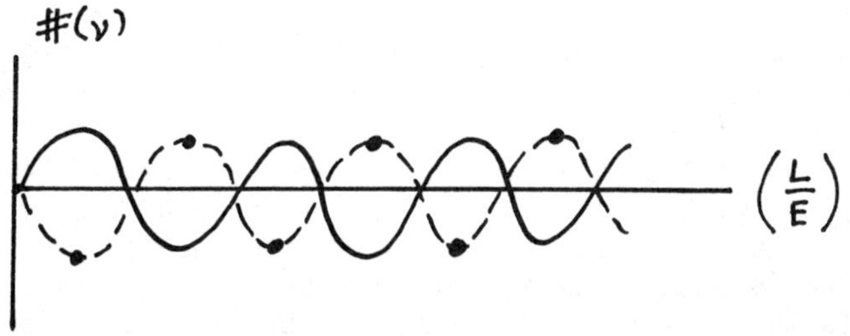

Fig. 4.5: Typical experimental scattered values. When fitting (4.30) to these points, it can be "in phase" (dotted line) or "out of phase" (solid line), depending on the values of Δm^2. Since the amplitude of the oscillations (4.30) is given by θ, clearly, the dotted line allows a much larger θ. One should therefore replace the oscillatory curve by a mean value line.

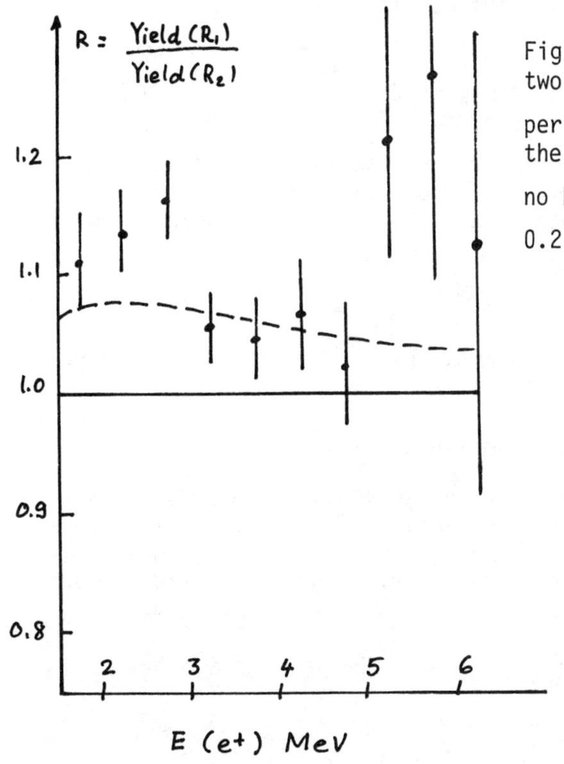

Fig. 4.6: Electron yields at two distances in the Bugey experiment.[43] The dashed line is the best fit to the two neutrino hypothesis; it gives $\Delta m^2 = 0.2\ eV^2$ $\sin^2\theta = 0.25$.

Recently, new experiments were done at Bugey. To increase statistics, the distances L were chosen small, 13.6 m and 18.3 m. This forces the detector to lie inside the reactor housing, making it harder to shield it completely.

Fig. 4.6 shows the ratio of the electron yields at the two distances, along with the best fit (4.35). The result is clearly incompatible with no oscillations, however the two-neutrino oscillation picture is not adequate. However, due to the difficulties mentioned, further results are hoped for. The allowed region of this experiment is plotted as a shaded region in Fig. 4.4. The results are compatible with Gösgen, but only, if one discards the reactor flux calculations completely.

b) Accelerator experiments

From table 4.1 we see that accelerator neutrinos, due to their larger energy, can produce heavier leptons. Therefore appearance experiments become possible.

Accelerator neutrino beams are usually dominated by muon neutrinos; therefore we expect results for $\nu_\mu \to \nu_e, \nu_\tau$ etc. The experimental bounds are given in Fig. 4.7.

A different arrangement is made in beam dump experiments,[49] such as at CERN. The high energy proton beam (400 GeV/c) is "dumped" into a target, producing heavy (charmed) particles. They decay quickly into neutrinos and other particles; one expects for them ("prompt neutrinos")

$$R = \frac{\nu_e + \bar{\nu}_e}{\nu_\mu + \bar{\nu}_\mu} \simeq 1 \qquad (4.36)$$

The slower decaying π, K, which produce mainly ν_μ are stopped before they decay; therefore at large distances one expects to see only the prompt neutrinos, and a deviation of R from 1 should signal oscillations.

These are a variety of results, however they have large errors (see Fig. 4.8)

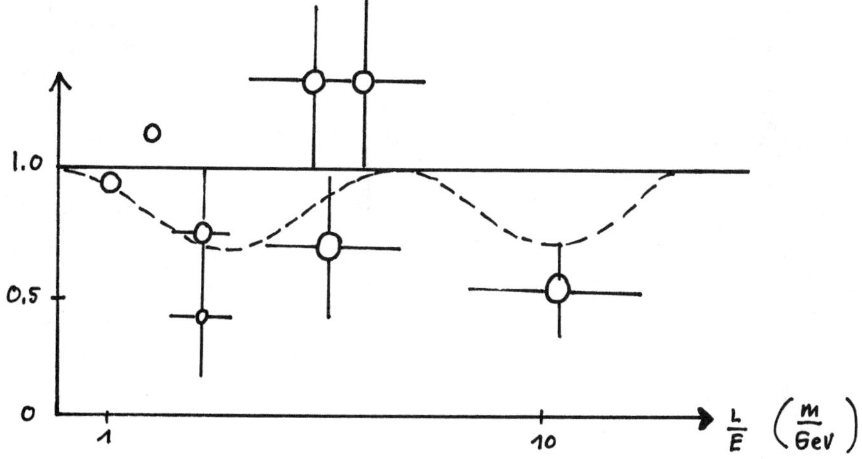

Fig. 4.8: R (4.36) plotted versus L = distance dump-detector. The dashed line denotes a recent fit by G. Conforto (Moriond Proceedings 1984) with $\Delta m^2 \overset{\sim}{=} 360\ eV^2$, $\sin^2 2\Theta \cong 0.32 \pm {0.18 \atop 0.08}$; hardly compatible with ref. 42) Θ is the angle between ν_e and ν_τ ; others seem small. The situation is uncertain; but it seems that these values could be ruled out by the ITEP experiment. Since due to the large Θ , the admixed neutrino could be visible; its mass would be $\lesssim 67$ eV, if $M_{\nu e} \lesssim 40$ eV .

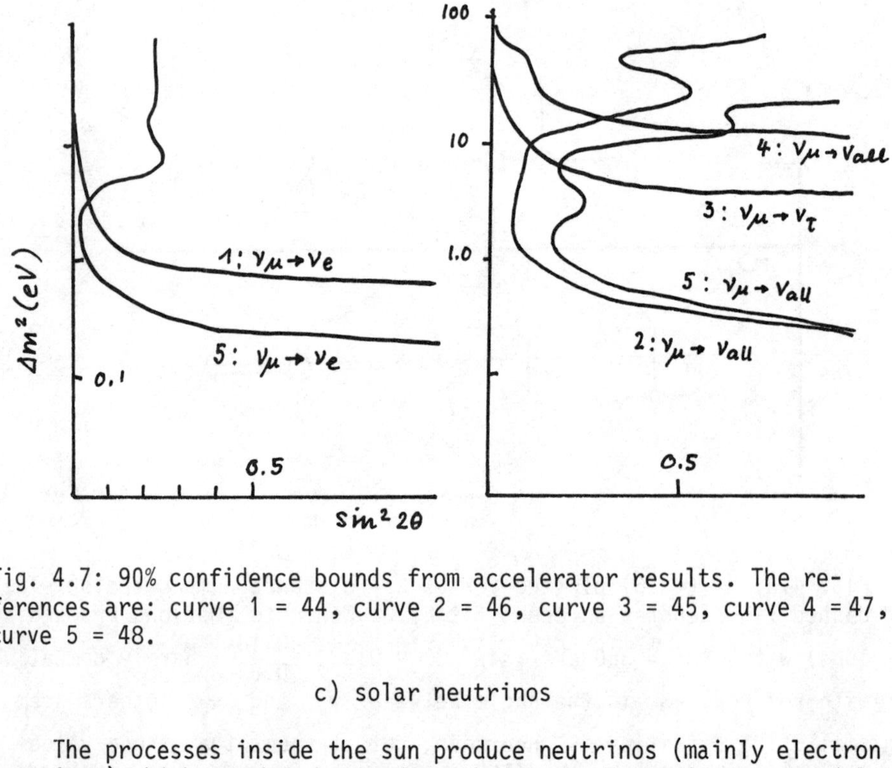

Fig. 4.7: 90% confidence bounds from accelerator results. The references are: curve 1 = 44, curve 2 = 46, curve 3 = 45, curve 4 = 47, curve 5 = 48.

c) solar neutrinos

The processes inside the sun produce neutrinos (mainly electron neutrinos) which can be observed on the earth. Depending on a model calculation of the sun's properties inside, a neutrino flux can be calculated and compared to the flux detected. In table 4.3 we give the main processes which produce neutrinos, their energy and expected numbers in two different detection experiments.

Reaction	Flux ($10^{10}cm^{-2}sec^{-1}$)	E_ν (MeV)	Cl	Ga
$p+p \rightarrow d+e^{+}+\nu$	6.1	0-0.42	0	67.2
$^{7}Be+e^{-} \rightarrow {}^{7}Li+\nu$	0.34	0.86 (90%) 0.34 (10%)	1.02	28.5
$^{8}B \rightarrow {}^{8}B + e^{+}+\nu$	6.0×10^{-4}	0-14	6.05	1.7
	Total (including all reactions)		7.6±3.3	$106\pm^{13}_{9}$

Table 4.3[50]: Some neutrino reactions in the sun. The last two numbers are the expected numbers in the ^{37}Cl and ^{71}Ga experiments. They are given in SNU (solar neutrino units = 10^{-36} captures/atom/second).

The pioneering experiment of Davis et al.[51] uses the detection reaction $\nu_e + {}^{37}Cl \rightarrow e^{-} + {}^{37}Ar$, with a threshold of 0.814 MeV. From table 4.3 we see that only few neutrinos, from the rarer reactions can be seen and that therefore the method is more dependent on solar model calculation. The advantage of using ^{37}Cl is that C_2Cl_4 is cheap and the reaction above can be detected relatively easily. The actual experiment used 3.8×10^5 liters of C_2Cl_4 ($\cong 10^{30}$Cl atoms) in a tank deep underground. The results are 2.1 ± 0.3 SNU.[51] Comparing to table 4.3 we have

$$R = \frac{\nu_e \text{ expected}}{\nu_e \text{ observed}} \simeq 3.2 \pm 1.5 \quad [52] \tag{4.37}$$

Since we do not expect that Δm^2 in (4.30) is so well adjusted that at L = distance (sun-earth) the oscillations just are very low*, we must interpret this as an average-effect e.g. very rapid oscillation (if we take $R \cong 3$) . As we have remarked, this implies then three species of neutrinos, maximum mixing angles, and Δm^2 10^{-11} eV^2 .[53] One test of this would be a neutral current reaction,

* Such a situation would also show a variation of R over the year (L changes) which has not been seen.

which should give $R = 1$. Although (4.37) is compatible with no oscillations, its general acceptance suffers from uncertainties in the model calculations in the sun;[54] it does not, however, contradict the Gösgen bounds.[42]

Table 4.3 indicates that a better detector would use a low threshold process. A possible reaction is $\nu + {}^{71}\mathrm{Ga} \rightarrow e^- + {}^{71}\mathrm{Ge}$[55] with threshold 0.236 MeV. No such experiment are completed; however sufficient amounts of Ga have been assembled.

Other neutrino oscillation manifestations are possible, for instance in cosmic ray neutrino experiments.[56]

d) Rare processes

If ν , ν mix , and have non-equal mass, the decay $\mu \rightarrow e\gamma$ is possible (Fig. 4.9),

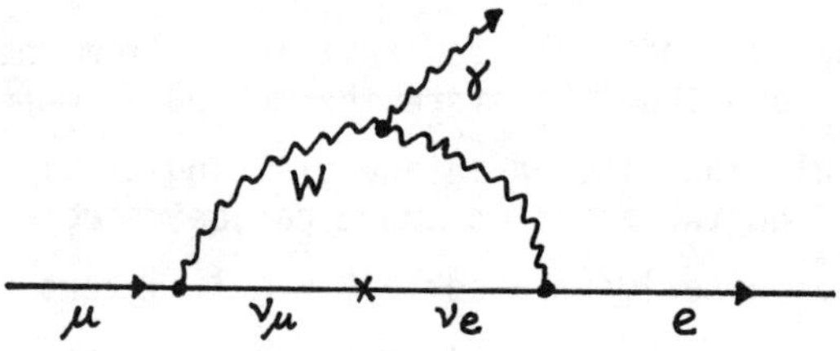

Fig. 4.9: Contribution to $\mu \rightarrow e\gamma$

One obtains[57]

$$R = \frac{\Gamma(\mu \rightarrow e\gamma)}{\Gamma(\mu \rightarrow \text{all})} = \frac{3\alpha}{32\pi}\left(\frac{\Delta m^2}{M_W^2}\right)^2 \sin^2\theta\cos^2\theta \qquad (4.38)$$

with the same notation as in (4.30). For $\theta = 45^{\circ}$, $R = 10^{-46}$. $[(\Delta m^2)^2$ in $\mathrm{eV}^4]$. Since $R_{\exp} < 10^{-11}$, this process is much less sensitive than neutrino oscillations.

e) Neutrino oscillations in matter[58]

So far, we have only discussed freely moving neutrinos. If neutrinos move through matter, the interaction of them with the surrounding can produce oscillation - like patterns, even if the neutrinos are massless. The requirement is, that not all neutrino types have the same neutral current interaction, or that the neutral current is not flavor diagonal - a situation discussed previously.

Consider the addition of a small interaction of the type

$$\delta H = (\cos\alpha(\bar{\nu}_1'\gamma_\mu\nu_1' + \bar{\nu}_2'\gamma_\mu\nu_2') + \sin\alpha(\bar{\nu}_1'\gamma_\mu\nu_2' + \bar{\nu}_2'\gamma_\mu\nu_1'))J^\mu(e,p,n) \tag{4.39}$$

where $J^u(p,n,e)$ is a current containing the particles through which the neutrinos ν_1, ν_2 travel. (4.39) can be rewritten (ν_1, ν_2 are taken massless)

$$\delta H = \cos\alpha(\bar{\nu}_1\gamma_\mu\nu_1 + \bar{\nu}_2\gamma_\mu\nu_2) + \sin\alpha(\bar{\nu}_1\gamma_\mu\nu_1 - \bar{\nu}_2\gamma_\mu\nu_2) \tag{4.40}$$

with

$$\nu_{1,2} = (\nu_1' \pm \nu_2')/\sqrt{2} \tag{4.41}$$

The states $\nu_{1,2}$ are eigenstates of the hamiltonian and develop as

$$\nu_i(x) = e^{-ikxn_i}\,\nu_i(o) \tag{4.42}$$

n being an "index of refraction", given by the interactions involved. Using (4.41, 4.42) we have

$$\begin{aligned} |\langle\nu_2'(0)|\nu_1'(x)\rangle|^2 &= \frac{1}{2}\left|e^{-ikxn_1} - e^{-ikxn_2}\right|^2 \\ &= 2(1-\cos kx(n_1-n_2)) = 4\sin^2\left[\frac{kx(n_1-n_2)}{2}\right] \end{aligned} \tag{4.43}$$

with

$$(n_1 - n_2) = \text{const} \times \sin\alpha \,. \tag{4.44}$$

The effects are only relatively large, if none of the ν_i' are ν_e, because of the charged current reactions $\nu_e e \to \nu_e e$, in the matter. Such experiments could be done with accelerator neutrinos (ν_μ) going through large amounts of earth. None have been carried out so far.

5. "KINEMATICAL" DETERMINATION OF NEUTRINO MASSES

The most direct test of neutrino mass is to simply detect it as a kinematical effect, using energy momentum conservation. Consider the decay of some particle X into a neutrino ν and some other known particles i. If their momenta and energies can be determined, the neutrino mass m is uniquely determined (unless there are oscillations, i.e. if the neutrino emitted is a linear combination of physical states, in which there are several solutions). Experimental considerations limit the particles i to be charged. The criteria for a successful measurement are:

a) the charged particles should have small kinetic energies, so that the resolutions Δp , ΔE are small
b) the 3-momentum of the neutrino should be small in order to enhance the effect of the mass
c) the decay satisfying a), b) should occur with sufficient branching ratio.

The simplest decay is the two-body decay $X \to \nu$ + lepton , where X is a meson, such as π , with a muon as lepton. Since E_{muon} P_{muon} P_ν $M_x - M_{lepton}$ is rather large, a) and b) are not satisfied; nevertheless there are experimental results. From the decay kinematics (ν and 1 move back to back), we have

$$|\vec{p}_e| = \frac{1}{2M_x}\sqrt{M_x^4 + M_\nu^4 + m_e^4 - 2M_x^2 m_\nu^2 - 2M_x^2 m_e^2 - 2m_\nu^2 m_e^2} \quad (5.1)$$

$$\simeq \frac{1}{2M_x}(M_x^2 - m_e^2) \quad ; \quad m_\nu \simeq 0 \quad (5.2)$$

and the error in m_ν due to $\Delta|\vec{p}_1|$ is

$$\Delta m_\nu^2 = \frac{M_x^2 - m_e^2}{M_x^2 + m_e^2} \, 2M_x \Delta |\vec{p}_e| . \quad (5.3)$$

(5.2) shows the rather unfortunate result that these experiments are only sensitive to the square of m_ν ; further (5.3) implies a large $\Delta m_\nu^2 \sim 2M_x\Delta|\vec{p}_1|$. With an SIN value[59] $\Delta|\vec{p}_{muon}| \cong 8.10^{-4}$ MeV , and $M_x = M_\pi \cong 140$ MeV , $\Delta m^2_{\nu_\mu}$ $8\cdot10^{-2}$ MeV and Δm_{ν_μ} 300 keV . Clearly this is hopeless for electrons.

In the actual experiment, π^+ are stopped and then decay into $\mu^+\nu_\mu$. Using (5.2) one obtains

$$m_{\nu_\mu} = (-0.163 \pm 0.082)\ \text{MeV} \tag{5.4}$$

giving an upper bound

$$m_{\nu_\mu} < 0.26 \quad \text{MeV} \tag{5.5}$$

on the mass of the μ -neutrino. The negative mean value in (5.4) has led to considering remeasuring m_π. Its statistical treatment requires some care.

If the muon neutrino is a linear superposition of mass eigenstates χ, e.g.

$$\nu_\mu = \sum_i{}' c_i \chi_i \tag{5.6}$$

then the decay rate into χ_i and muon is[60]

$$\Gamma = \frac{G_F^2 f_\pi^2}{8\pi^2} M_\pi m_\mu^2 \left(1 - \frac{m_\mu^2}{M_\pi^2}\right)^2 \times \frac{|c_i|^2\, 2 M_\pi\, |\vec{p}_\mu|}{m_\mu^2 (M_\pi^2 - m_\mu^2)^2} \left\{ M_\pi^2 m_\mu^2 + M_\pi^2 m_{\chi_i}^2 - (m_\mu^2 - m_{\chi_i}^2)^2 \right\} \tag{5.7}$$

and we expect to find indication for the χ_i by peaks in the $|\vec{p}_\mu|$ distribution, at values given by (5.1). Such searches have been made, the results being negative. The limit are displayed in Figs. 5.1 and 5.2.

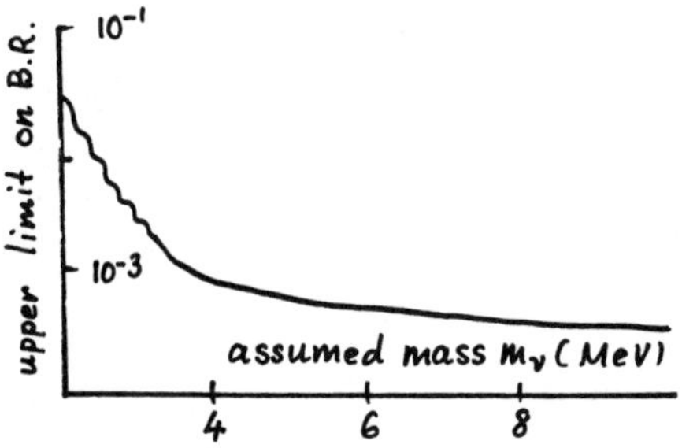

Fig. 5.1: The upper limit for the branching ratio for pion decay to a muon and a massive neutrino. The results are compared to those of ref (62) (dots). (From ref. 61)

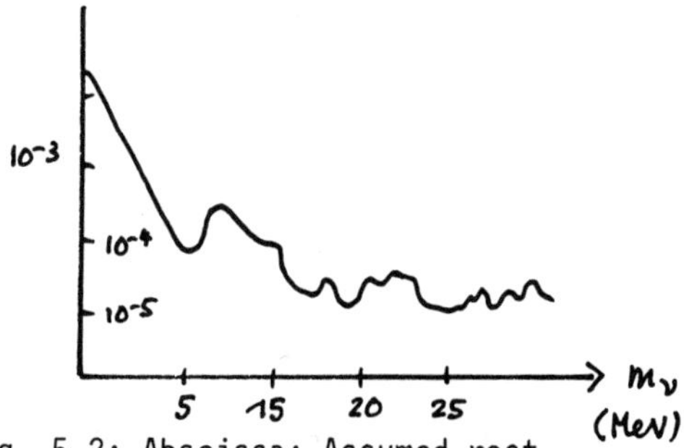

Fig. 5.2: Abscissa: Assumed rest mass m_ν of a hypothetical heavy neutrino ν_H emitted in the decay $\pi^+ \to \mu^+\nu_H$. Ordinate: Upper limit for the branching ratio $\eta = \Gamma(\pi^+\to\mu^+\nu_H)/\Gamma(\pi^+\to\mu^+\nu_\mu)$: ν_μ is the light neutrino.

(From ref. 62)

The three body decay, although at first sight more complicated, proves to be better suited for the mass determination. In $X \to Y + l + \nu_l$, the energy spectrum of the lepton l is continous, (which led Pauli to postulate the neutrino); the maximal energy is given by

$$E_{max}(\ell) = M_X^2 + m_e^2 - (M_Y + m_\nu)^2 \tag{5.8}$$

We note the linear dependence on m_ν in (5.8).

The energy spectrum of the lepton is given by[63]

$$\frac{dN(E)}{dp} = F_c(Z)\, p^2 (Q-E)\sqrt{(Q-E)^2 - m_\nu^2} \tag{5.9}$$

where F_c is the "Coulomb" factor and Q is the energy which is released in the decay (the energy difference of X and Y). The common way to plot dN/dp is to divide out $F(Z)p^2$ and take the square root (Kurie plot)[63]

$$K = \sqrt{\frac{dN(E)}{dp \cdot F_c \cdot p^2}} = (Q-E)\sqrt{(Q-E)^2 - m_\nu^2} \tag{5.10}$$

$$\simeq (Q-E) - \frac{m_\nu^2}{(Q-E)} \qquad (m_\nu \approx 0) \; .$$

For $m_\nu = 0$ this is just a straight line, but for $m_\nu \neq 0$ but small the line drops (Fig. 5.3). For various neutrinos coupled to 1 , there are several "dents" in the Kurie plot. However, finite energy resolution tends to deform the ideal curve (Fig. 5.4)[64].

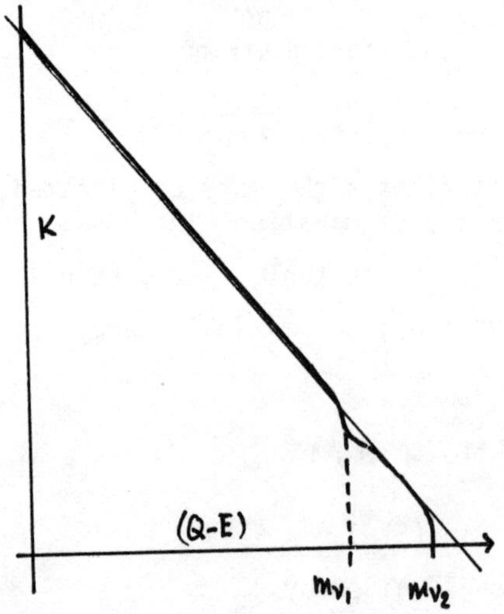

Fig. 5.3: ideal Kurie plots for $m_\nu = 0$ and two non-zero values for m_ν

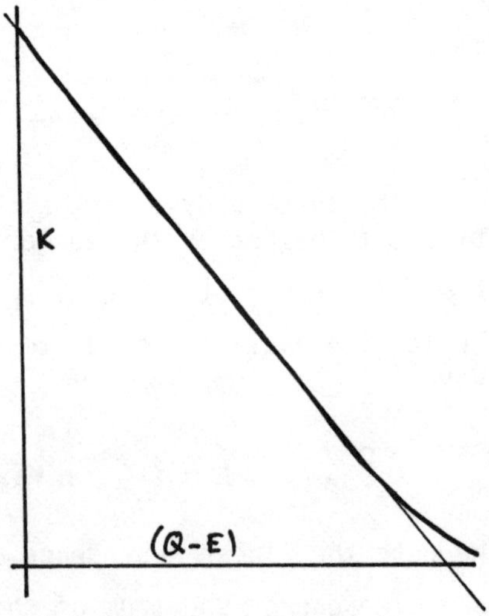

Fig. 5.4: effect of the finite energy resolution on the end-point energy.

A particularly good system is the tritium decay[65]

$$^3H \rightarrow {}^3He + e^- + \bar{\nu} \qquad (\bar{\nu} = \nu_{RCe}) \tag{5.11}$$

It has a low Q value of 18.6 keV, a relatively manageable atomic structure and a lifetime $t_{1/2} \sim 12$ years. Early attempts gave $m < 500$ eV ;[65] 1972 Bergkvist[64] improved there results dramatically and obtained $m_\nu < 55$ eV , we give the newest numbers shortly.[66]

At the new level of accuracy all complications must be carefully considered. Some of them are:

a) The initial tritium has one bound electron $\sim 1_s$ (E = 13.6 eV) (if it is a free atom). When one of the neutrons in the nucleus decays, the electron, which does not participate can either stay in a Is state or be "shaken up" to 2s state in the final ^{3}He , at a 43 eV higher energy; in principle to all other states, including the continuum. It is estimated that the probabilities for the ^{3}He to be in the states 1_s , 2_s , others, are 70%, 25%, 5%, respectively. The effect of this widening of the energy resolution as illustrated in Fig. 5.5.

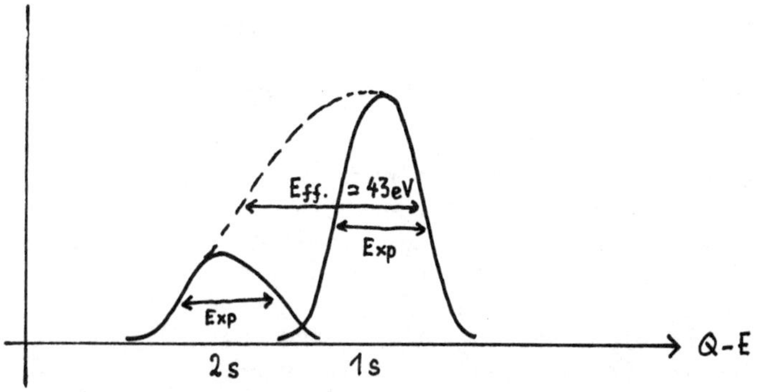

Fig. 5.5: Overlap of the two experimental line forms producing an effective, much wider curve. Estimates give a broadening from 55 to 70 eV[64] for the experiment in ref. 64.

b) Tritium is bound in some material. In the ITEP experiment it is a complicated molecule, $C_5H_{11}NO_2$ (Valine) whose levels also are shaken and changed during the decay. (The 11 hydrogen atoms can be 9 ordinary and 2 tritium)

c) Background is serious. It can be eliminated by accelerating electrons coming from the ^{3}H by a fixed, known amount (last reference of 66)

d) line form: The detector "sees" the monoenergetic electron not like a gaussian (as in Fig. 5.5) but in some asymmetric form. In the newest ITEP experiment (ref. 66) the detector is strongly scrutinized.

The experimental results from the ITEP experiment are given in table 5.1 and Fig. 5.6. They indicate a non-vanishing electron neutrino mass,

$$20\ eV < m_\nu < 45\ eV \tag{5.11}$$

This range of values is somewhat high when considered from the point of view of grand unified models discussed previously; models with intermediate mass scales may, however, give such values (3.7). As we will see, cosmological considerations point to masses in the range (5.11) if neutrinos are to play an important role in the history of the universe. Clearly, further experiments are needed to confirm or disprove (5.11).

	Valine	Atom	Triton (Nucleus)
M_ν	34.8 ± 1.9	30.9 ± 1.6	$13.8^{+2.5}_{-3.5}$
E_0	18584.2 ± 1.6	18580.5 ± 1.3	18567.4 ± 1.0
Δm	18603.6 ± 6	18608 ± 6	18586.8 ± 6

Table 5.1: Calculated effects of various models for the tritium β decay (last of ref. 66). In the first column, the energy levels of the whole molecule are included, in the second only the atomic ones (such as in Fig. 5.5) and in the last column only the effect of the nucleus was included. E_0 is the endpoint energy of the electron, Δm the T-^3He mass difference. The latter has been precisely determined very recently (see above ref.) to be 18599 ± 3 and thus agrees extremely well with the table. Three measurements have been done for three thicknesses of the Tritium source and agree with each other. From the table we see that even "worst" case (Triton) gives $m > 9$ eV .

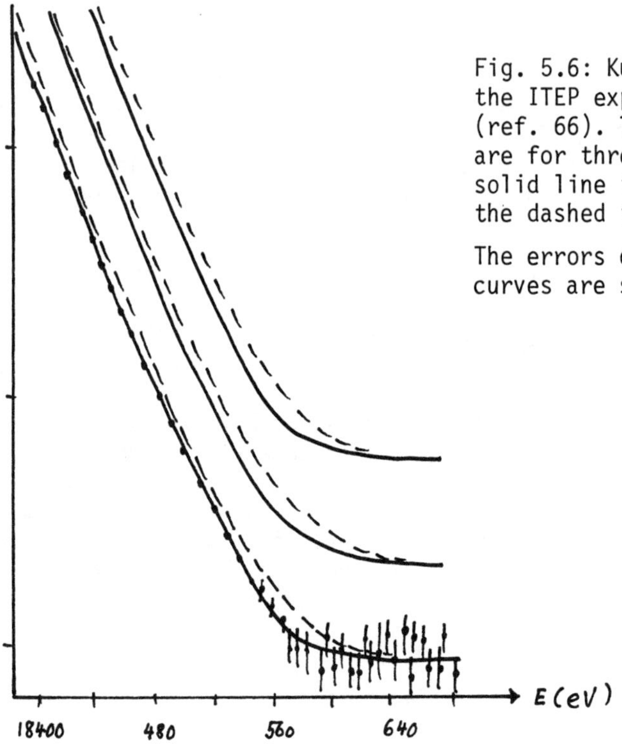

Fig. 5.6: Kurie Plot for the ITEP experiment (ref. 66). The three curves are for three targets. The solid line is the best fit; the dashed is for $m_\nu = 0$. The errors on all three curves are similar.

Various further processes have been proposed or used to determine m_ν :

International Bremsstrahlung:[67] A neutron deficient nucleus can capture an electron from its electron shell, thereby emitting a neutrino. The electron leaves a "hole" which is filled when a higher lying electron falls in it, emitting energy (X Ray or Auger electron). Thus the process is $(A,Z) \to (A,Z-1) + \nu_e + \gamma$ and can be used to determine m if adequate atoms are available. Using ^{193}Pt and ^{163}Ho , a bound of 1.3 keV has been obtained.[68]

Three body decays, such as $\Pi \to \nu_\mu \mu \gamma$.[69] The advantage of these decays has been discussed. Due to the low rate the experiments are difficult. No data exists.

The decay $K \to \mu\, \nu_i$ was[70] used to determine admixtures of other massive neutrinos, as discussed before.

The decay $K \to \Pi\, \mu\, \nu_\mu$ [71] was used to give an upper limit on $m(\nu_\mu)$. $\Pi \to e\, \nu$ decay[72] was considered.

$K \to e\, \gamma$ [73] was considered.

6. DOUBLE β -DECAY

If ordinary β-decay $(A,Z) \to (A,Z+1)$ is impossible or strongly supressed, then (A,Z) might decay through two successsive β -decays into $(A,Z+2)$, or through some new non-renormalizable interaction.[74] In the spirit of renormalizable gauge theories, we only consider the first possibility. We can then distinguish between

$$(A,Z) \to (A,Z+2) + e^- + e^- + \bar{\nu}_e + \bar{\nu}_e \tag{6.1.1}$$

$$(A,Z) \to (A,Z+2) + e^- + e^- \tag{6.1.2}$$

The two reactions can be distinguished by observing the number of electrons vs. their total energy. In case (6.1.2) the total energy is fixed by the energy difference of (A,Z) and $(A,Z+2)$, in case (6.1.1) the distribution is spread out since the neutrinos can share in the available energy. Processes (6.1) can dominate the ordinary β -decay in some even-even nuclei; the best known examples are

$$\begin{aligned} {}^{58}\mathrm{Ni} &\to {}^{58}\mathrm{Fe} \\ {}^{48}\mathrm{Ca} &\to {}^{48}\mathrm{Ti} \\ {}^{130}\mathrm{Te} &\to {}^{130}\mathrm{Te} \\ {}^{128}\mathrm{Te} &\to {}^{128}\mathrm{Te} \\ {}^{82}\mathrm{Se} &\to {}^{82}\mathrm{Kr} \end{aligned} \tag{6.2}$$

where the transitions are from the 0^+ to 0^+ or 2^+ excited states.

Further "double weak" processes can be written down; for example

$$\begin{aligned} (A,Z) &\to (A,Z-2) + e^+ + e^+ \\ e + (A,Z) &\to (A,Z-2) + e^+ \\ K^+ &\to \pi^- + \mu^+ + \mu^+ \\ \Sigma^- &\to p + \mu^- + \mu^- \end{aligned} \tag{6.3}$$

The processes (6.1.2), (6.3) violate lepton number by two units. As we have seen, Majorana masses have this property. Therefore, these processes can be used to search for them. Thus, unlike the previous mass tests, "neutrinoless" double β -decay is sensitive only to a certain kind of mass.

Using the hypothesis of the iterative weak interaction we can analyse the lepton-number violating process, as in Fig. 6.1. In general, the W_i can be both "left- and right-handed". We write

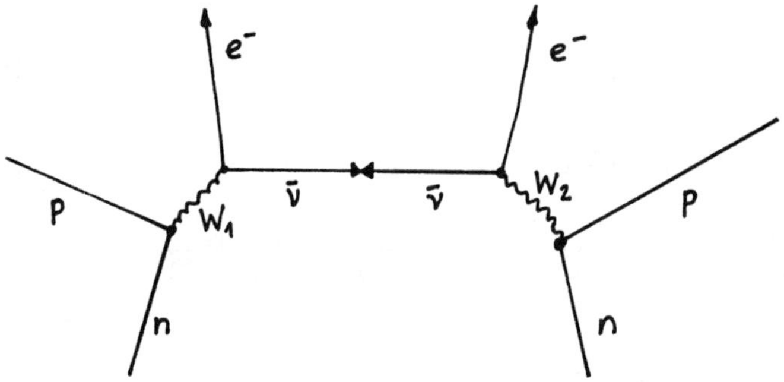

Fig. 6.1: Neutrinoless double decay through two weak processes and neutrino annihilation

for their interactions with the leptons (a = b = real)

$$g_1 W_1^{\mu} (\bar{e} \gamma_{\mu} (a_1 L + b_1 R) \nu) \tag{6.4}$$

$$g_2 W_2^{\mu} (\bar{e} \gamma_{\mu} (a_2 L + b_2 R) \nu) \tag{6.5}$$

The amplitude for Fig. 6.1 is then proportional to

$$(\bar{e} \gamma_{\mu} (a_1 \nu_L + b_1 \nu_R)) (\bar{e} \gamma_{\nu} (a_2 \nu_L + b_2 \nu_R)) \tag{6.6}$$

To evaluate (6.6) we rewrite the second term in (6.6) to bring ν_L, ν_R into a form such that the propagators (2.36) or (2.37) can be used. We also need to rewrite ν_L, ν_R, ν_{LC}, ν_{RC} in terms of mass eigenstates:

$$\nu_L = (A^+ X_{(L)} + C^+ X_{(R)})_L \tag{6.7.1}$$

$$\nu_R = (B^T S^*_{(L)} X_{(L)} + D^T S^*_{(R)} X_{(R)})_R \tag{6.7.2}$$

$$\nu_{LC} = (B^{+}\chi_{(L)} + D^{+}\chi_{(R)})_L \tag{6.7.3}$$

$$\nu_{RC} = (A^{T}S^{*}_{(L)}\chi_{(L)} + C^{T}S^{*}_{(R)}\chi_{(R)})_R \tag{6.7.4}$$

now,

$$\bar{e}\gamma_\nu(a_2\nu_L + b_2\nu_R) = -(a_2\nu_L{}^T + b_2\nu_R^T)\gamma_\nu{}^T\bar{e}^T \tag{6.8.1}$$

$$= (a_2\bar{\nu}_{RC} + b_2\bar{\nu}_{LC})\gamma_\nu e_C \tag{6.8.2}$$

Then, the double β-amplitude is proportional to:
i) Form (6.8.1) (use (4.5.2) and (6.7.1), (6.7.2))

$$\begin{aligned}&\langle [a_1(A^{+}\chi_{(L)} + C^{+}\chi_{(R)})_L + \\ &\quad b_1(B^{T}S^{*}_{(L)}\chi_{(L)} + D^{T}S^{*}\chi^{T}_{(R)})_R] \times \\ &\times [a_2(A^{+}\chi^{T}_{(L)} + C^{+}\chi^{T}_{(R)})_L + \\ &\quad b_2(B^{T}S^{*}{}_{(L)}\chi^{T}{}_{(L)} + D^{T}S^{*}\chi^{T}_{(R)})_R]\rangle\end{aligned} \tag{6.9}$$

$$\begin{aligned}&= a_1a_2\left(\frac{A^{+}\hat{C}\,S\,m_{(L)}A^{+}}{p^2 - m^2_{(L)}} + \frac{C^{+}\hat{C}\,S\,m_{(R)}C^{+}}{p^2 - m^2_{(R)}}\right) \\ &+ a_1b_1\left(\frac{A^{+}\hat{C}\not{p}\,B^{T}}{p^2 - m^2_{(L)}} + \frac{C^{+}\hat{C}\not{p}\,D^{T}}{p^2 - m^2_{(R)}}\right) \\ &+ \ldots\end{aligned} \tag{6.10}$$

Here $\hat{C}$ is the charge conjugation operator; the hat is to distinguish it from the matrix C, p is the momentum of the neutrinos. In (6.10), matrix indices are omitted for simplicity; the first term means, for example $A^{+}{}_{1k}\,\hat{C}\,m_{(L)k}\,A^{+}{}_{1'k} \cong \langle\nu_{L1}, \nu_{L1'}\rangle$.

ii) Form (6.8.2) (use (4.5.2) and (6.7.2) and (6.7.4).) We get now

$$\langle (a_1 \nu_L + b_1 \nu_R)(a_2 \bar{\nu}_{RC} + b_2 \bar{\nu}_{LC}) \rangle$$

$$= a_1 a_2 \frac{A^+ m_{(L)} S A^+}{p^2 - m^2_{(L)}} + \cdots \tag{6.11}$$

and we see that the two expressions coincide.

In the absence of right-handed currents, $b_1 = b_2 = 0$, $a_1 = a_2 = 1$; then from (6.10) we get that the amplitude for omitting two leptons 1 , 1' is proportional to

$$\varkappa^0_{LL} = \sum_i \frac{A^+_{\ell i} A^+_{\ell' i} S_{(L)i} m_{(L)i}}{p^2 - m^2_{(L)i}} + \sum_i \frac{C^+_{\ell i} C^+_{\ell' i} S_{(R)i} m_{(R)i}}{p^2 - m^2_{(R)i}} \tag{6.12}$$

If there are also right-handed currents, the amplitude $\varkappa^0_{RR}$ is analogous to (6.12); if $W_1 = W_L$ $W_2 = W_R$, we get, using the $a_1 b_2$ term in (6.10)

$$\varkappa^0_{LR} = \sum_i \frac{A^+_{\ell i} B^T_{\ell' i} \not{p}}{p^2 - m_i^2} + \sum_i \frac{C^+_{\ell i} D^T_{\ell' i} \not{p}}{p^2 - m_i^2} \tag{6.13}$$

---

* In principle, due to $W_L - W_R$ mixing, there are several amplitudes proportional to $\varkappa^0_{LR}$, etc. For instance, there are 4 ampli-

Let us discuss some of the consequences of (6.12) and (6.13). (6.12) is explicitely dependent on the masses m_i and is thus very small for small masses (the masses in (6.12, 6.13) are of course the positive masses (2.)), whereas (6.13) contains the masses only indirectly, through the mixing matrices. It will, however, also be accompanied by a suppression factor m_W^2/m_{WR}^2 since it belongs to a diagram with a W_R -exchange.

$\mathcal{H}^0_{LL}$ contains the factors S_i . This has the consequence that if some S_i are negative, the various terms in (6.12) can cancel against each other.[77] This is most clearly seen in case of pure Dirac neutrinos, where (4.9), (4.10)

$$C = A \qquad m_{(R)i} = m_{(L)i} \qquad S_{(L)i} = - S_{(R)i} \tag{6.14}$$

and thus $\mathcal{H}^0_{LL} = 0$. This is the result one expects because in this case there is no lepton number violation.** But also in case of pure Majorana masses this can happen. Take as example only 2 states ψ_1, ψ_2 with a mass matrix

$$M_L = \begin{pmatrix} a & c \\ c & b \end{pmatrix}$$

$$\frac{1}{2}\,tg\,2\alpha = \frac{c}{a-b}\,;\quad 2\cdot\lambda_{1,2} = (a+b) \pm \sqrt{(a-b)^2 + 4c}$$

---

* tudes proportional to $\mathcal{H}^0_{LR}$, depending on the possible helecities of the quarks. See refs. 83, 78 for details.

** This is analogous to the following well known case: The intermediate gauge bosons, $W^\pm$, have a definite charge. If we write them as $W^\pm = \frac{W_1 \pm iW_2}{2}$ then the $W_{1,2}$ seemingly violate charge conservation; but since any process has contributions from both W_1 and W_2 , the charge violating amplitude cancel.

$$U^T M_L U = \begin{pmatrix} \lambda_1 & 0 \\ 0 & \lambda_2 \end{pmatrix}$$

$$\mathcal{U} = \begin{pmatrix} \cos\alpha & -\sin\alpha \\ \sin\alpha & \cos\alpha \end{pmatrix}$$

Using (6.12) one gets $(S_i = \lambda_i/|\lambda_i|)$

$$(\mathcal{H}^{\circ}_{LL})_{11} = \cos^2\alpha\, |\lambda_1|\, S_1 + \sin^2\alpha\, |\lambda_2|\, S_2$$

which can clearly vanish, for instance, if $a = b = 0$ (Pseudo-Dirac fields).[78]

In many models we considered ((3.4), (3.5)) there is a relatively small Dirac term, say m, and a large Majorana term M for the right-handed neutrinos. Then

$$\mathcal{O}(B) \approx \mathcal{O}(C) \simeq \frac{m}{M} \ll 1 . \qquad (6.15)$$

Since $p^2 \approx O(1\ \text{MeV})$ (nuclear energy), $M \gg p^2$ and $m^2 \ll p^2$

$$\mathcal{H}^{\circ}_{LL} \approx \frac{A^+A^+ S_{(L)}\, m_{(L)}}{p^2} + \left(\frac{m}{M}\right)^2 \frac{\mathcal{U}^+\mathcal{U}^+ S_{(R)}}{M} \qquad (6.16)$$

where U is defined by $C = \frac{m}{M}U$ and $O(U) \approx 1$.

Similarly, in these models

$$\mathcal{H}^{\circ}_{LR} \cong \left(\frac{m}{M}\right)\left(\frac{A^+ V^T p}{p^2} + \frac{{}^+ D^T \not{p}}{M^2}\right) \qquad (6.17)$$

where $B = \frac{m}{M}V$. Since $\frac{m}{M} \approx \frac{m_W}{m_{W_R}}$, there will be an extra factor $\left(\frac{m}{M}\right)^2$

In addition to the above mechanisms involving W , one can initiate neutrinoless (ββ) -decay by fancy scalars, such as in Fig. 6.2.

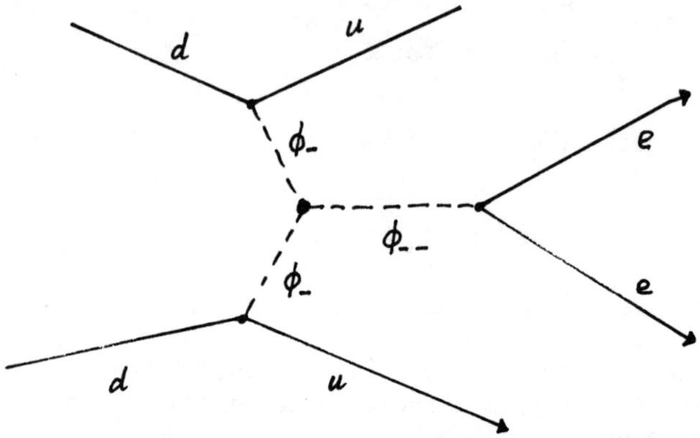

Fig. 6.2: Neutrinoless double β -decay via Higgs processes.[79] This mechanism was shown[80] to be small.

Another possibility is through emission of a light boson (Fig. 6.3)

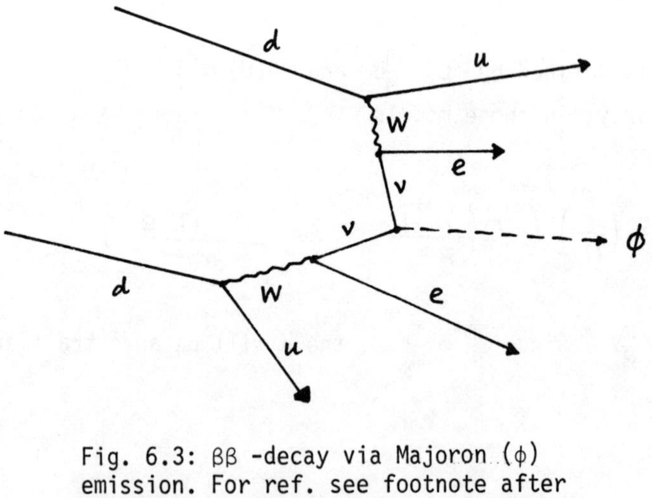

Fig. 6.3: ββ -decay via Majoron (ϕ) emission. For ref. see footnote after Eq. (3.2). In this case, the 2-electron spectrum differs from (6.1.2); still another possibility is shown in Fig. 6.4.

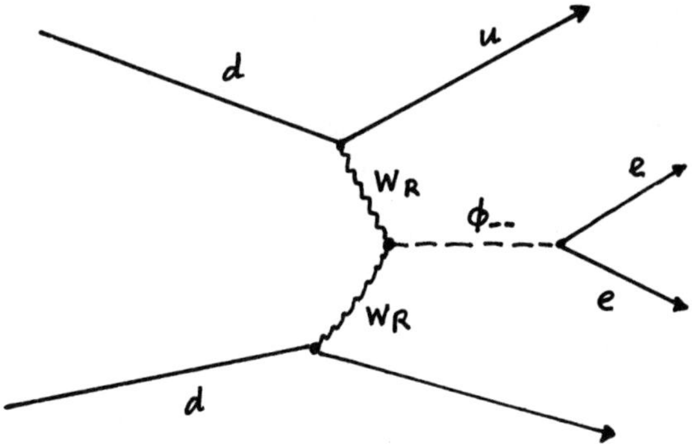

Fig. 6.4: Neutrinoless double β -decay via W_R and Higgs processes. Only W_R could give appreciable contributions.[81]

In case of heavy Majorana neutrinos, one can also have a contribution from Fig. 6.5.[82]

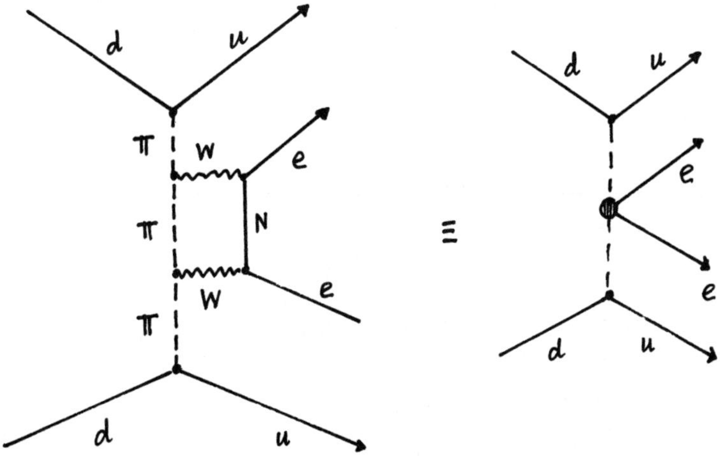

Fig. 6.5: Contribution from heavy Majorana neutrino. The corresponding lepton violating parameter is similar to the previous ones, but the calculation of the matrix-element is different.

Having now recognized the lepton number violating processes, we must calculate the corresponding matrix-elements to obtain the decay rates or lifetimes. An elaborate program is that of the Osaka group;[83] further calculation are described in refs. 86. We will just mention a few items.

Since one wants two β -decays, they can either take place from two different neutrons or from one nucleon, if it is an excited state of negative charge[84] (Figs. 6.6)

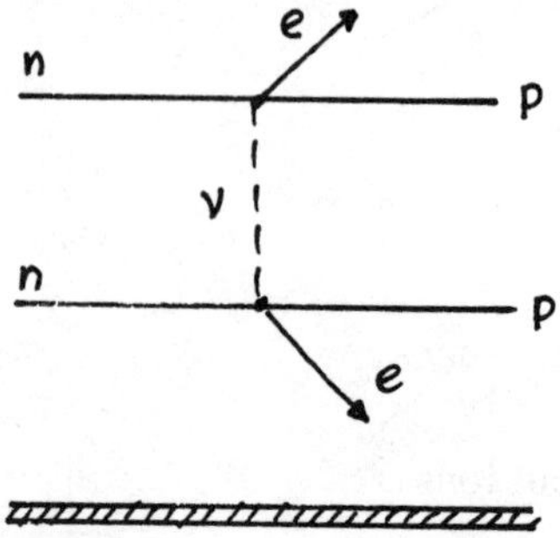

Fig. 6.6.1: 2 N mechanism

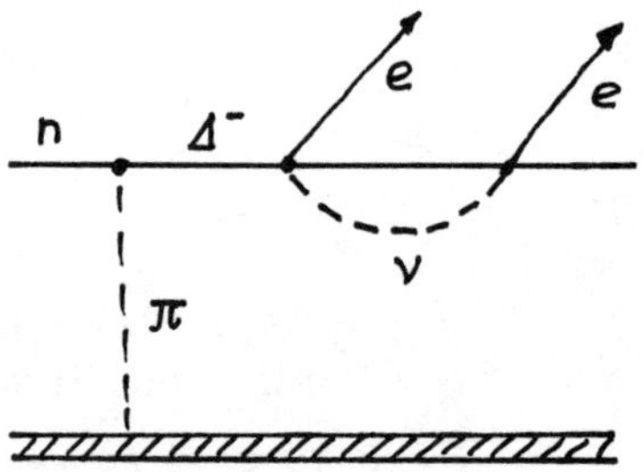

Fig. 6.6.1: N* mechanism

There are $0^+ \to 0^+$ transitions and $0^+ \to 2^+$ transitions, possibly distinguishable experimentally. Depending on the lepton violating process, some of the amplitudes are forbidden (Table 5.1)

	$\mathcal{H}^0_{LL,RR}$		$\mathcal{H}^0_{LR}$	
	2N	N*	2N	N*
$0^+ \to 0^+$	yes	no	yes	no
$0^+ \to 0^+$	no	no	yes	no

Table 6.1:[83] Selection rules for ββ -amplitudes.

The decay rate is now given by[83]

$$\Gamma \simeq |\alpha\eta_{LL} + \beta\eta_{LR} + \gamma\eta_{RR}|^2 \tag{6.18.1}$$

$$= (C_1\eta_{LL} + C_2\eta_{LL}\eta_{RR} + \dots)\,|M|^2 \tag{6.18.2}$$

where the $\eta_{LL}, \ldots$ are derived from (6.12), (6.13) (see 6.16) and (6.17)):*

$$\eta_{LL} = \sum_i \frac{A_{ei} A_{ei} S_i m_i}{m_e} \equiv \frac{\langle m \rangle}{m_e} \tag{6.19}$$

$$\eta_{LR} = \left(\frac{m_{W_L}}{m_{W_R}}\right)^2 \sum_i A^+_{ei} B^T_{ei} \ .$$

$|M|^2$ is the nuclear (Gamov-Teller) matrix element. The parenthesis in (6.18.2) is not an absolute square because the various terms in (6.18.1) involve other spin configurations. The factors C_i are given in ref. 83). g^2 is the gauge coupling. We see, however, that any experiment will put a limit on a combination of η_{LL} and η_{LR}; (6.18.1) suggests the dependence to be about an ellipse (see Fig. 6.7).

One usually distinguishes the experiments into: 1) geochemical experiments. In these, one analyses old ores of a known age (typically $\sim 10^9$ years) which have a lot of the parents in (6.2), and one searches for the products of (6.2).

The best measured cases are

$$\begin{aligned} {}^{130}\mathrm{Te} &\to {}^{130}\mathrm{Xe} \\ {}^{128}\mathrm{Te} &\to {}^{128}\mathrm{Xe} \\ {}^{82}\mathrm{Se} &\to {}^{82}\mathrm{Kr} \end{aligned} \tag{6.20}$$

In such experiments, only the lifetime being measured, one cannot directly distinguish between 2ν or 0ν double β-decay. However, if measured values exceed calculated ones for 2ν decay, one might infer the 0ν decay.

Double β-decay has been seen by this method.[85] The average values are

---

* For heavy neutrinos (6.16) it is useful to define η by $\frac{C^+C^+S\,m}{M}$ with $m \cong 0.85$ GeV[81]

$$T_{1/2}\ (^{82}Se) = (1.45 \pm 0.15)10^{20}Y$$
$$T_{1/2}\ (^{130}Te) = (2.55 \pm 0.2)\ 10^{21}Y \qquad (6.21)$$

Theoretical calculations[86] yield, however (2ν)

$$T_{1/2}^{theory} \Big/ T_{1/2}^{Exp} \cong \begin{cases} \frac{1}{4} & ^{82}Se \\ \frac{1}{100} & ^{130}Te \end{cases} \qquad (6.22)$$

which shows the uncertainties which still plague the calculations. The theoretical value gives a shorter lifetime.

A somewhat calculation independent quantity is[75]

$$(R_T)^{-1} = \frac{\Gamma(^{128}Te)}{\Gamma(^{130}Te)} = \frac{\Gamma^{2\nu}(1+\Gamma^{0\nu}/\Gamma^{2\nu})\ (^{128}Te)}{\Gamma^{2\nu}(1+\Gamma^{0\nu}/\Gamma^{2\nu})(^{130}Te)} \qquad (6.23)$$

The analysis in ref. 83 gives the ratios in (6.23).[87] Assuming that they are reliable, measuring $(R\ )^{-1}$ gives us information on $\Gamma^{0\nu}$. There are two measurements:

$$(R\ )^{-1} = (0.6 \pm 0.005)10^{-3} \quad ^{88} \qquad (6.24.1)$$

$$(R\ )^{-1} = (1.03 \pm 1.13)10^{-4} \quad ^{85} \qquad (6.24.2)$$

while the first is incompatible with non-zero $\Gamma^{0\nu}$, the second is depending on the analysis one gets $\langle m \rangle < 4.7$; $\eta_{LR} <$ $\eta_{LR} < 2.0\ 10^{-5}$ (ref. 83), for (6.24.2). Additional measurements seem required.

2) Laboratory experiments. In these experiments, a material which can undergo double β -decay is put in a well shielded place (a tunnel in the mountains) and one searches for the two omitted electrons. The best investigated system is ^{76}Ge -decay[89] into ^{76}Se . Ge is well suited as source as well as electron detector. The Milan group has the best bound on the 0ν - ββ -decay lifetime:[90]

$$T_{1/2}^{0\nu} > 1.2\ \ 10^{23}\ Years$$

further limits are given in table (6.1)

UCB/LBL[91] :	3.6	10^{22} Y
Guelph [92] :	3.2	10^{22} Y
Osaka [93] :	2.2	10^{22} Y
Calteck[94] :	2.0	10^{22} Y
Batelle[95] :	1.3	10^{22} Y

Table 6.1: Present limits on the double β-0ν decay lifetime for various experiments

Fig. 6.7 gives the bounds on <m> (6.19) and η_{LR} .

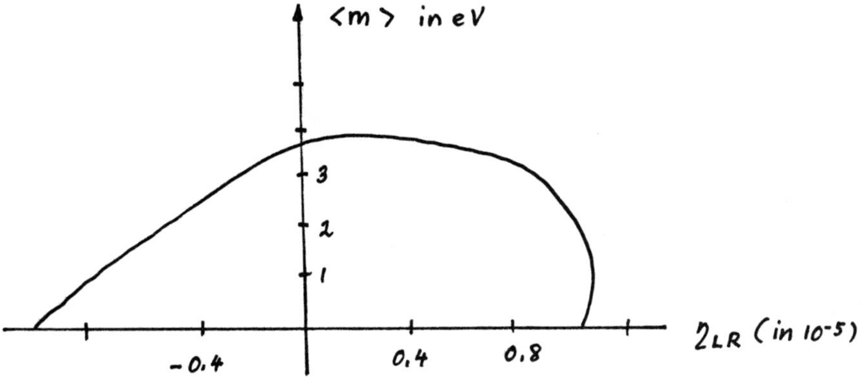

Fig. 6.7: Bounds on <m> and η_{LR} from the ^{76}Ge experiment[90], using the calculations of ref. 86. As remarked (6.22), the calculation might be uncertain; a factor 4 (6.22) would allow a mass ≅ 10 eV , compatible with ITEP experiment. Similar bounds are obtained, if the W_L - W_R mixing is included.[86,90]

7. NEUTRINOS AND THE UNIVERSE

There is a growing connection between elementary particle physics and astrophysics. In this chapter we discuss some of these connections relating to neutrinos.[96]

In order to have ready a few formulas needed later, we briefly describe the standard cosmology.[97]

In an isotropic and homogeneous universe (we assume it to be filled with a "fluid") we can write for the metric (Robertson-Walker)

$$ds^2 = dt^2 - R^2(t)\, d\vec{s}^{\,2} \tag{7.1.1}$$

with

$$d\vec{s}^{\,2} = \frac{dr}{1-kr^2} + r^2(d\theta^2 + \sin^2\theta\, d\varphi^2) \tag{7.1.2}$$

in spherical coordinates. $k = 0\ ,\ \pm 1$ is the curvature. Using (7.1), Einsteins Equations $R_{\mu\nu} - \frac{1}{2} g_{\mu\nu} R - (\lambda g_{\mu\nu}) = -8\pi G\, T_{\mu\nu}$ and the form

$$T_{\mu\nu} = p g_{\mu\nu} + (p+\rho)\,\delta_{o\mu}\delta_{o\nu} \tag{7.2}$$

one obtains, for $\lambda = 0$

$$3\ddot{R} = -4\pi G(\rho + 3p) R \tag{7.3.1}$$

$$R\ddot{R} + 2\dot{R}^2 + 2k = 4\pi G(\rho - p) R^2 \tag{7.3.2}$$

where $\dot{R} = \frac{dR}{dt}$ and G is Newtons constant

$$G = \frac{0.74}{10^{38}\,\text{GeV}} = \frac{1}{M^2_{\text{Planck}}} \tag{7.4}$$

$$M_{Pl} = 1.2\ 10^{19}\ \text{GeV} = 2\cdot 10^{-5}\ g$$

(7.3.1) and (7.3.2) yield

$$\dot{R}^2 + k = \frac{8\pi G}{3} \rho R^2 \tag{7.5}$$

together with the equation of energy conservation

$$\dot{p} R^3 = \frac{d}{dt}\left(R^3 \cdot (p + \rho)\right)$$
$$\frac{d}{dR}(\rho R^3) = -3pR^2 \tag{7.6}$$

(7.5) and (7.6) can be solved if $\rho(R)$ is known. Such models are called Friedmann models.

Before we proceed, we note that the red-shift of light emitted at t_1 and arriving at t_0 is

$$Z = \frac{R(t_0)}{R(t_1)} - 1 \tag{7.7}$$

If $R(t_0) > R(t_1)$, $Z > 1$: red shift which is observed; thus $R/R > 0$ at present. As long as $p + \rho > 0$, $R/R < 0$, and therefore R as a function t must be curved downward as we go back in time (Fig. 7.1).

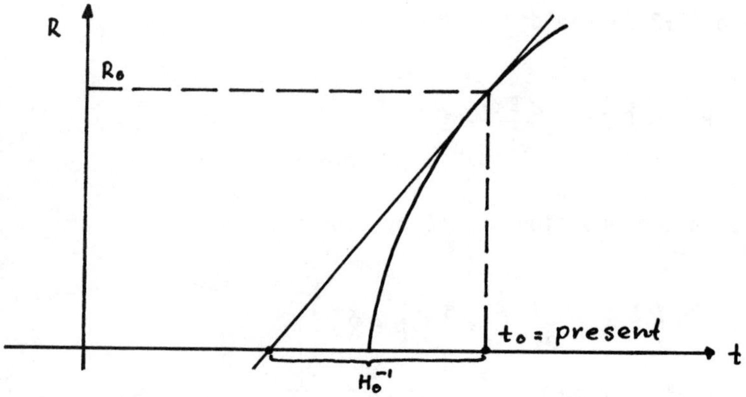

Fig. 7.1: R as a function of time t for $\ddot{R} < 0$, $\dot{R} > 0$ as explained in the text. H_0^{-1} is the Hubble time. It is larger than the age of the universe.

As $t \to \infty$, (7.6) says that $(p > 0)$ ρ decreases faster than R^{-3}. Thus, from (7.5) we learn that

$$\left.\begin{array}{lll} k=-1 & \dot{R}^2 > 0 & R \to \infty \\ k=0 & \dot{R}^2 \to 0 & R \to \infty \end{array}\right\} \text{expansion}$$

$$\left.\begin{array}{l} k=1 \quad \dot{R} = 0 \quad \text{for} \quad \rho R^2 = \dfrac{3}{8\pi G} \\ \quad \text{but later} \quad \dot{R} < 0 \quad (\ddot{R} < 0) \end{array}\right\} \text{contraction}$$

(7.5) gives $(\dot{R}/R = H = \text{Hubble's constant})$ at present $(t = t_0,\ R = R_0$ etc.)

$$\rho_0 = \frac{3H_0^2}{8\pi G} + \frac{3k}{8\pi G R_0^2} \tag{7.8}$$

There are two measurements of H_0 ; writing

$$H_0 = \varkappa \cdot 50 \text{ km/sec Mpc} \quad (\text{Mpc} = 3.3 \cdot 10^6 \text{ Light years}) \tag{7.9}$$

$$\varkappa = 1 \ \text{ref. 98} \qquad \varkappa = 2 \ \text{ref. 99}$$

(The Hubble time is $\sim \frac{R}{\varkappa} 10^{10}$ Y)

The critical density ρ_c is that for $k = 0$;

$$\rho_c = 0.5 \quad 10^{-29} \varkappa \ g/cm^3 \tag{7.10}$$

such that

$$\frac{\rho_0}{\rho_c} = \left(1 + \frac{k}{H_0^2 R_0^2}\right) = 2q_0^{*} \tag{7.11}$$

where

$$q = -\frac{R\ddot{R}}{\dot{R}^2} \tag{7.12}$$

is the decceleration parameter. It can be marginally measured; we will only use a very conservative limit of $q_0 < 1.5$ although indications exist that $q_0 < 0.5$.[100] Using (7.10) we get

$$\rho_0 \lesssim 1.5 \cdot 10^{-29} \ g/cm^3 \tag{7.13}$$

We can solve (7.5) for two special cases. At high temperatures ($T > 1$ MeV) , (radiation dominated universe) $p = \rho/3$, and (7.6) gives

$$\rho \sim R^{-4} \tag{7.14}$$

In the matter dominated era (non relativistic particles) $\rho \ll \rho$, thus, from (7.6))

$$\rho \sim R^{-4} \tag{7.15}$$

With (7.5) and $K = 0$, we have (7.14)

---

* $p = 0$: $\ddot{R} R/\dot{R}^2 = -4\Pi g\rho/H^2$ and (7.3.1): $\ddot{R}R/\dot{R}^2 + 2(1+k/\dot{R}^2) = 4\Pi g\rho/H^2$

$$\frac{\dot{R}}{R} = \sqrt{\frac{8\pi G}{3}}\ \frac{1}{R^2} \qquad \text{or}$$

$$t(R) = \frac{1}{2}\sqrt{\frac{3}{8\pi G}}\ R^2 \cong \frac{1}{2}\sqrt{\frac{3}{8\pi G}}\ \frac{1}{T^2} \tag{7.16}$$

where T(R) is the time elapsed until the scale is R . The last equation in (7.16) follows from (7.20), see below.

Another ingredient are the formulas for the number densities for particles in thermal equilibrium. We have, at an equilibrium temperature T *

$$n(T)_{\substack{\text{Fermi}\\ \text{Bose}}} = \frac{g}{2\pi^2}\int_0^\infty \frac{p^2\,dp}{\exp(E/T)\pm 1} \tag{7.17}$$

where g = number of degrees of freedom (2 per photon, 2 per neutrino handedness) $E^2 = p^2 + m^2$. If m = 0 , we can integrate (7.17) and get (ζ = Riemann- ζ-function)

$$n(T)_{\substack{B\\ F}} = g\frac{T^3}{\pi^2}\begin{cases}\zeta(3)\\ \frac{3}{4}\zeta(3)\end{cases} = g\frac{T^3}{\pi^2}\begin{cases}1.22\\ 0.91\end{cases} \tag{7.18}$$

Equally, the energy density is given by (m = 0)

$$\rho(T)_{\substack{B\\ F}} = \frac{g}{2\pi^2}\int\frac{p^3\,dp}{\exp(E/T)\pm 1} = \frac{3gT^4}{\pi^2}\begin{cases}\zeta(4)\\ \frac{7}{8}\zeta(4)\end{cases} = \begin{cases}0.33\ gT^4\\ 0.28\ gT^4\end{cases} \tag{7.19}$$

* The "complete" form is

$$n(T) = \left(\frac{g}{2\pi^2}\right)\left(\frac{kT}{\hbar c}\right)^3\int\frac{dz\ z^2}{\exp[(E-\mu)/kT]\pm 1} \qquad z = pc/kT,\quad E^2 = (pc)^2 + (mc^2)^2$$

From (7.14) we then have (radiation domination)

$$R \sim 1/T \tag{7.20}$$

7.1. Upper limits on the neutrino mass

In the previous sections we saw that the neutrino's mass, if non-zero, is small. However, even such a small mass may affect (gravitationally) the history of the universe, if there is a large number of neutrinos.[101] In the following we will calculate this effect in the framework of the hot big bang model of the universe.

In the early universe, when temperatures were very high*, all light particles were in thermal equilibrium, through processes like

$$e^+ + e^- \longleftrightarrow \nu + \bar{\nu} \quad ^{**} \tag{7.21.1}$$

$$e^+ + e^- \longleftrightarrow \gamma \tag{7.21.2}$$

$$\gamma \longleftrightarrow \nu + \bar{\nu} \tag{7.21.3}$$

The density of neutrinos can be obtained by considering in detail the reactions (7.21). Since for a temperature $T \gg m_e$, (7.21.2) produces enough e^+e^- to bring $\nu, \bar{\nu}$ into thermal equilibrium by (7.21.1).

The rate of producing ν through(7.21.1) is for $T \ll M_W$,

$$\sigma \sim (G_F^2 T^2) T^3 \tag{7.22}$$

where G_F is the Fermi constant. The factor in parantheses is the weak cross section where the average energy of the external particles is assumed to be purely thermal kinetic motion, and the factor T^3 counts the number of electrons (Eq. 7.18)

---

* It is customary to use units where Boltzmanns constant is 1. then 1^0 Kelvin = $0.85\ 10^{-4}$ eV ; 1 MeV = $1.2 \cdot 10^{10}\ {}^0$K.

** ν stands here for all light neutral leptons.

This means, that, per second, a neutrino makes σ annihilation (with another neutrino), or that it can travel $1/\sigma$ freely. On the other hand, the horizon length at time t is $ct = t$. If $1/\sigma > t$, then the neutrino travels freely through the universe, and the thermal equilibrium cannot establish itself any more*. Using (7.16) we have

$$\begin{aligned} G_F^2 T^5 &\simeq T^2/M_P^2 \\ T &\simeq \left(\frac{1}{M_P G_F^2}\right)^{1/3} \simeq 1\ \mathrm{MeV} \simeq 10^{10}\ {}^\circ\mathrm{K} \end{aligned} \tag{7.23}$$

For lower temperatures, the neutrinos decouple. This analysis depended on the assumption $m_\nu \ll 1$ MeV .

The particle number densities are (7.18):

$$\begin{aligned} n_\nu &= 0.091\ g T^3 \quad ^{**} \\ n_\gamma &= 0.122\ g T^3 \end{aligned} \tag{7.24}$$

Thus, assuming the number of neutrinos to remain constant since $T = T_\nu$,*** we get for their density:

$$n_\nu\ (T = T_{present}) = n_\nu\ (T_\nu) \left(\frac{T_{pr.}}{T_\nu}\right)^3 \tag{7.25}$$

As to the photons, the above assumption does not hold, since at $T \cong m_e = 0.5$ MeV, electrons annihilate into photons, but not vice versa (electron decoupling). Then, we get extra photons as follows.

---

* Any interaction outside the horizon does not affect this (causality).

** Numerically, $n_\nu(2.7^0\mathrm{K}) \cong 300\ \nu/\mathrm{cm}^3$; the baryon density is about $\sim 10^{-6}\ \mathrm{B/cm}^3$; this shows that if $m_\nu > M_p\ 10^{-8}$, neutrinos dominate.

*** This means that no neutrinos are decaying etc.

In a comoving volume $V \sim R^3$, the entropy S is approximately conserved (neglecting non-adiabatic processes). The entropy is

$$S = \frac{4}{3}\frac{\rho}{T}V = VT^3\frac{4}{3}\left(g_B + g_F\cdot\frac{7}{8}\right)1.35 \qquad (7.26)$$

Thus we get, for photons and electrons

$$\begin{aligned} S_\gamma + S_e &= \frac{4}{3}VT^3\left(2 + \frac{7}{8}4\right)1.35 \; * \\ &= 2.7\left(\frac{11}{4}\right)\frac{4}{3}VT^3 \end{aligned} \qquad (7.27)$$

Thus, we see that after the annihilation $e^+e^- \to \gamma$ the effective temperature for photons is $\left(T\sqrt[3]{\frac{11}{4}}\right)$. This yields

$$\frac{n_\nu}{n_\gamma} = \frac{3}{4}\cdot\frac{4}{11}\cdot\frac{g}{2} = \frac{3}{11} \qquad (7.28)$$

for each neutrino species with $g = 2$ (one handedness).

To obtain an upper mass limit, we use equation (7.13). The density due to neutrinos ν_i of mass m_i must satisfy (7.13):

$$2\sum_i m_i n_{\nu_i}^{**} = \sum m_i \frac{3}{11} n_\gamma < 1.5\cdot 10^{-29} g/cm^3$$

Thus

$$\sum m_i \lesssim 40\ eV \quad \text{ref. 101, 102} \qquad (7.29)$$

where $n_\gamma \sim 400/cm^3$, and 10^{-24} g ≅ 0.5 eV. The bound (7.29) is close to the experimental limit (ref. 66).

---

* The electron and antielectron have together $g = 4$.

** The factor 2 is to count ν and $\bar{\nu}$.

If neutrinos are heavier than 1 MeV, they are non-relativistic when they decouple.* We use therefore a distribution[103]

$$n_\nu = 2T^3 \left(\frac{m_\nu}{2\pi T}\right)^{3/2} e^{-m_\nu/T} \tag{7.30}$$

instead of (7.18). The rate of the weak annihilation rate is now ($E \gtrsim m_\nu$)

$$\sigma \simeq G_F^2 m_\nu^2 n_\nu x_s \tag{7.31}$$

where x_s counts the channels $\nu\bar{\nu} \to x$.
Now the expansion rate $\dot{R}/R$ is (7.1), (7.5)

$$\frac{\dot{R}}{R} = \sqrt{\frac{8\pi G \varrho}{3}} \tag{7.32}$$

Here, c counts the light ($m < T$) particles, $c \geqq 1.5$.

As before, we set the rates (7.31) and (7.32) equal, to obtain the decoupling temperature.
Defining $y = m_\nu/T$, we get[104]

$$1 = \frac{2\, G_F^2 x_s e^{-y} y^{1/2}}{(2\pi)^{3/2}} \sqrt{\frac{3}{8\pi G c}} \left(\frac{m_\nu}{m_{Prot}}\right)^3 m_{Prot}^3 \tag{7.33}$$

or

$$1 \simeq 10^7 x_s\, y^{1/2} e^{-y} \left(\frac{m_\nu}{m_{Prot}}\right)^3 \frac{1}{\sqrt{c}}$$

---

* We anticipate this.

If $m_\nu \cong m$ (proton) then $y \gtrsim 7 \cdot \ln 10 + \ln (7 \cdot \ln 10 \frac{x_s}{\sqrt{C}})$. Taking $x_s = 14$ $(\nu_e, \nu_\mu, e, u, d, s)$, then $y \cong 20$, or a temperature of 50 MeV.

From (7.30) and (7.18) one has

$$\frac{n_\nu}{n_\gamma} = e^{-y} y^{3/2} \frac{0.5}{(11/4)} \simeq 4.5 \; 10^{-8} \tag{7.34}$$

where the factor (11/4) takes into account the rise in the photon temperature (7.27). This formula is correct for all temperatures (below 0.5 MeV).

Replacing $y^{3/2} e^{-y}$ by (7.33) one gets

$$n_\nu = n_\gamma \; 10^{-7} \frac{2}{11} \frac{\sqrt{C}}{x_s} y \left(\frac{m_{Prot}}{m_\nu}\right)^3 \tag{7.35}$$

or

$$2 \sum_i m_i \nu_i = \sum m_i \frac{4}{11} \frac{\sqrt{C}}{x_s} n_\gamma \left(\frac{m_{Prot}}{m_\nu}\right)^3 < 1.5 \; 10^{-29} \; g/cm^3 \tag{7.36}$$

$$\sum_i \left(\frac{m_{\nu_i}}{m_{Prot}}\right)^2 \frac{x_s}{\sqrt{C}} \gtrsim 33 \tag{7.37}$$

For one neutrino, this gives

$$m_\nu \gtrsim 2 \text{ GeV} \tag{7.38}$$

However, for several neutrinos, the bound is weaker and decreases with $1/\sqrt{\text{number of } \nu}$

7.2 Bounds on the number of light neutrinos

We give only a crude description of this estimate from the helium abundance.[105]

For temperature $T \geqq T_{crit}$, protons and neutrons are in equilibrium, and

$$\frac{n_{Neutron}}{n_{Proton}} \simeq e^{-(m_{Neut.} - m_{Prot.})/T} \qquad (7.38.1)$$

T_{crit} is determined, as before, by setting the expansion rate and the rate for weak processes equal to each other. According to (7.5) and (7.19), the expansion rate is proportional to the number of light particles, n_L :

$$\frac{\dot{R}}{R} = \text{const.}\; n_L\, T^2$$

Using (7.12) for the weak reaction rate,

$$T_{crit} \sim n_L^{1/3} \qquad (7.39)$$

Below T_{crit}, the numbers of N,P remain essentially constant. As most of the N are inside ^{4}He nuclei, a large T_{crit} means a larger ^{4}He abundance, by Eq.(7.38.1). From measurements, this can be obtained, and using (7.38.1) a bound on n_L is obtained. In this way one gets

$$\text{number of extra neutrinos} \lesssim 3 \qquad (7.40)$$

This coincides with limits from measurements of the Z -width[106] and leads to speculations that there are indeed only three generations of fermions.

In the non-relativistic limit, for instance $v_s^2 \to 0.2 \langle v^2 \rangle = 0.6 \frac{p}{\rho}$.
M_j is essentially the mass of a bulk of fluid which can "clump" together by gravitational attraction overcoming the thermal motion. Smaller masses cannot and "diffuse" again.
For ideal fermi gases, pressure and density are (Eqs. 7.18, 7.19)

$$p = \frac{4g\pi}{3h^3} \int_0^\infty \frac{dp\, p^4}{(\exp \beta\epsilon + 1)\epsilon} \qquad \epsilon^2 = m^2 + p^2 \tag{7.46}$$

$$\rho = \frac{4g\pi}{h^3} \int_0^\infty \frac{\epsilon p^2\, dp}{(\exp \beta\epsilon + 1)}$$

With this

$$M_J = \frac{4\pi}{3} \left(\frac{\pi}{G}\right)^{3/2} \frac{p^{3/2}}{\rho^2} . \tag{7.47}$$

To make this more transparent, one may introduce

$$x = \frac{m_\nu}{T} \simeq \frac{1 + z_{T=m_\nu}}{1+z}$$

and gets[109]

$$M_J = \frac{4\pi}{3} \left(\frac{\pi}{G}\right)^{3/2} \left(\frac{h}{8\pi}\right)^{1/2} \left(\frac{1}{3}\right)^{3/2} g(x)\, x^2 \simeq$$

$$\simeq \frac{1.77\ M_{PL}^3 \times g(x)}{m_\nu^2}$$

$$g(x) = \left(\int_0^\infty \frac{y^4 dy}{(e^y+1)(y^2+x^2)^{1/2}} \right)^{3/2} \left(\int \frac{y^2 dy\, (y^2+x^2)^{1/2}}{(e^y+1)} \right)^{-2}$$

$$y = \frac{p}{T} \qquad \epsilon \simeq p \tag{7.48}$$

We see from this that if $m_\nu > 30$ eV , neutrinos can "condensate" under their gravitational force to form the "basis" for galactic clusters. M_J is the order of M_H .

Another way to see which neutrino masses can form cluster is as follows. Consider a relatively dense neutrino gas in a region as big as a galaxy, about 10^2 kpc . Due to the exclusion principle, the number of neutrinos is limited, because neutrinos of a velocity larger than V_{esc}

$$V_{esc} = \sqrt{G M_{c\ell}\, 2 / R_{c\ell}} \; * \tag{7.49}$$

will escape. Here $M_{c\ell}$ is the total mass of the neutrinos, and $R_{c\ell}$ is the radius of the (spherical) region we consider. If the neutrinos are not heavy enough, then their total mass will therefore not equal the desired cluster mass, $\sim 10^{15}\ M_{SUN}$.

The maximum number of neutrinos is (by the Pauli principle**)

---

* Since $\frac{m_\nu V^2_{esp}}{2} = \frac{m_\nu M_{c\ell}\, G}{R c\ell}$ defines the escape velocity

** Phase space cell $\Delta x \Delta p = h$.

$$n_{max} = 2n \int_0^{m_\nu V_{esc}} \frac{dp\, p^2}{8\hbar^3 \pi^3} (4\pi) = \frac{n}{3\hbar^3 \pi^2} \left(m_\nu V_{esc}\right)^3 \qquad (7.50)$$

Now we set $M_{c\ell} = n_\nu m_\nu \frac{4\pi}{3} R^3_{ce}$; this gives

$$M_{c\ell} = n_\nu m_\nu \frac{4\pi}{3} R^3_{c\ell} < \frac{4n\, m_\nu^4 G^{3/2} M_{c\ell}^{3/2} 2^{3/2} R^3_{c\ell}}{9\hbar^3 \pi R_{c\ell}^{3/2}} \qquad (7.51)$$

$$m_\nu^8 > \frac{81\pi^2 \hbar^6}{2^7 n^2 G^3 M_{c\ell} R^3_{c\ell}} \qquad (7.52)$$

where n is the number of fields with 2 helicity states.
Taking $n = 2$, $(\nu_e, \bar{\nu}_e)$; $R_{c\ell} = 100$ kpc, $M = 10^{15}\, M_{SUN} \cong 10^{48}$ g one gets

$$m_\nu \gtrsim 3\,eV \qquad (7.53)$$

This treatment, although approximate, differs little from a complete calculation of the self consistent equilibrium of neutrinos under gravity. These one gets, instead of $81\pi^2/n^2\, 2^7$, $81\pi^2/n^2\, 2^3$ or a factor of $\sqrt{2}$ for m_ν.[111]

Another very interesting approach uses the properties of a non-interacting fluid (Vlasov equation)[110]. It is more general than above treatment and gives a slightly (factor 2 1/4) sharper bound.

Numerical calculations[112] show that galaxy formation with neutrinos of mass $\simeq$ 30 eV occurs too late. One possibility to solve the dilemma (see Eq. (7.29) is to consider more massive neutrinos which then decay after "their work is done".

7.4 Unstable neutrinos

Once neutrinos are massive, they can in principle decay*. Owing to their low masses, decay rates are presumably small and the decay may not be visible in experiments; however, on a cosmic scale, effects might be detectable. Unstable neutrinos might solve the problems previously mentioned. If heavy enough, they can effectively help forming galaxies but must decay fast enough not to upset ρ/ρ_{crit}.

Here I will not treat unstable neutrinos in detail, but summarize a few results.

Neutrinos decay through $\nu \to \nu' + B$, $\nu \to \nu'\nu'\nu'$ etc., where B is a boson, the photon or another light particle (Majoron, ...).**

---

* We will concentrate here on light neutrinos, say of a mass less than m_e. Heavy neutral leptons will most certainly decay.

** Se References in the footnote after eq. (3.2)

The simplest decay which would occur in most models with unstable neutrinos is $\nu \to \nu' + \gamma$. The decay would proceed similarly to $\mu \to e\gamma$ (ref. 57). The photon energy is $(m_\nu^2 - m_{\nu'}^2)/2m_\nu$, limits are obtained from not observing such photons.

As an example consider the solar neutrinos. They have a flux $\sim 10^{11}\,cm^{-2}\,sec^{-1}$ and an average energy of 0.2 MeV. The flight to the earth is about 500 sec. If t_L is the decay time in the sun-earth-coordinate system, then

$$\begin{aligned} N_\nu &\sim N_0\, e^{-t/t_L} \\ N_\gamma &\sim N_0\, (1 - e^{-t/t_L}) \end{aligned} \tag{7.54}$$

Observations limit the photon flux to be less than $10^{-4}\,cm^{-2}\,sec^{-1}$.[113] Therefore

$$N_\gamma = 10^{11}(1 - e^{-500\,sec/t_L}) < 10^{-4} \tag{7.55}$$

One is thus justified to expand the exponent and gets

$$t_L > 5 \cdot 10^{17}\,sec \tag{7.56}$$

Converting into the neutrino system (t_o):

$$t_0 = \frac{m_\nu}{E_\nu} t_L > m_\nu(eV)\ 10^{12}\,sec \tag{7.57}$$

This can be compared to (4.38) for $\nu \to e\gamma$ (the exact calculation involves of course more diagrams (photon from internal fermion line)). Replacing $m_\mu \to m_\nu$, $m_N \to m_\mu$, one gets

$$\Gamma(\nu \to \nu'\gamma) \simeq \frac{3\alpha}{32\pi}\left(\frac{m_\mu}{M_W}\right)^4 \sin^2\theta \cos^2\theta\ 10^6 \left(\frac{M_\nu}{m_\mu}\right)^5 \simeq 10^{-41}\,sec^{-1} \quad (30\,eV \hat{=} m_\nu) \tag{7.58}$$

The strongest of such bounds come from absence from photons in white dwarf formation[113] and supernovae explosions[114]; the result is

$$\begin{aligned} \tau_0 &> m_\nu(eV)\ 10^{17}\,sec \\ 0.1\,MeV &\leq E_\nu \leq 10\,MeV \end{aligned} \tag{7.59}$$

In the cosmological context, $\nu \to \nu' + \gamma$ yields photons which should be included in the 2.7 degree background.

If the lifetime τ_L is comparable to the lifetime of the universe t_o or longer, the rate of ν decays per cm^3, n_ν/τ_L is roughly a constant, and the total number density of photons is $(n_\nu/\tau_L)t_o$. The flux is thus $(n_\nu/\tau_L)t_oC$. Since, roughly, $t_oC \simeq R_{UNIV}$, the expected flux of photons is

$$\sim n_\nu \cdot R_{UNIV}/\tau_L \tag{7.60}$$

where n_ν is the present neutrino density, $n_\nu \sim 400/cm^3$.

If $m_\nu \simeq 10^{-3}$ eV, the photons could distort the microwave background; their flux is limited by $\sim 10^{12}$. This gives

$$\tau_L \gtrsim 400\, R_{UNIV}\, 10^{-12} \simeq 10^{18}\, sec \qquad (7.61)$$

If $m_\nu \simeq 1$ eV, the photons are optical and bound by a flux $\simeq 10^8$. Then

$$\tau_L \gtrsim 400\, R_{UNIV}\, 10^{-8} \simeq 10^{22}\, sec \qquad (7.62)$$

Thus, it is likely that if a neutrino decays into photons its lifetime exceeds the lifetime of the universe, and it cannot serve to initiate galaxies.*)

Neutrino decays can also be bound from the requirement that the fermions ν_L produced in $\nu \to \nu_L + \ldots$ do not have an energy density which exceeds ρ_c. As an example, take $\nu \to \nu_L + \gamma$. The energy of ν_L (taken massless) is $m_\nu/2$.

At the decoupling time t_D (when $\nu_L + \gamma \to \nu$ is inoperative to restore equilibrium), the total number of ν is $N = n_\nu(t_D)\, \frac{4\pi}{3} R^3(t_D)$. If the decay rate is Γ, then, per second, $\Gamma N e^{-\Gamma t}$ ν_L are produced. Their energy, at time t, is $\frac{m}{2}$, but is red-shifted to $\frac{m}{2}\left(\frac{t}{t_0}\right)^{1/2}$ at time t = today. Therefore, the total contribution of the ν_L to the present density is

$$\rho(\nu_L, t_0) = \frac{1}{\frac{4\pi}{3} R^3(t_0)}\, \frac{m}{2} N \int \Gamma e^{-\Gamma t} \left(\frac{t}{t_0}\right)^{1/2} dt \qquad (7.63)$$

Since $R(t_0)/R(t_D) = \sqrt{t_0/t_D}$, we get

$$\rho(\nu_L, t_0) = n_\nu(t_D)\, \frac{m}{2}\, \Gamma \left(\frac{1.9}{T(t_D)}\right)^3 \int e^{-\Gamma t} \sqrt{\frac{t}{t_0}}\, dt \qquad (7.64)$$

where we have included the usual factor $\sqrt[3]{\frac{4}{11}}$ for the temperature.

Requiring (7.64) to be less than $\rho_{crit} \simeq 1.5\ 10^{-29}\,\mathrm{cm/cm^3}$, we get, taking $\Gamma^{-1} \gg t_D$

$$\frac{m_\nu}{2}\, n_\nu(t_D) \left(\frac{1.9}{T(t_D)}\right)^3 \frac{1}{(t_0\Gamma)^{1/2}} \int_0^{t_0\Gamma} e^{-\rho} \sqrt{\rho}\, d\rho < \rho_{crit} \qquad (7.65)$$

This can be compared directly to the case of stable neutrinos where we had $2m_\nu\, n_\nu(t_D) \left(\frac{1.9}{T(t_D)}\right)^3 < \rho_{crit}$. Using the two different forms for n_ν (7.24) and (7.36) we get **)

$$m_\nu \ll 1\,\mathrm{MeV}:\quad m_\nu \lesssim \frac{80\,\mathrm{eV}}{g(t_0\Gamma)} = \frac{80\,\mathrm{eV}}{\left(\frac{1}{t_0\Gamma}\right)^{1/2} \int_0^{t_0\Gamma} e^{-\rho}\sqrt{\rho}\, d\rho} \qquad (7.66)$$

---

*) There are also effects from photon reheating etc.[116]

**) $T(t_0)$ is taken ~ 1 MeV; it depends on the decay mechanism.

$$\left(\frac{m_\nu}{m_{Prot}}\right)^2 \frac{X_s}{\sqrt{c}} \gtrsim 33.4 \cdot g(\Gamma t_o) \qquad m_\nu \gtrsim 1\,\text{MeV} \tag{7.67}$$

If $t_o \gg \Gamma^{-1}$, then g can be approximated by $(\frac{1}{t_o\Gamma})^{1/2}\,\Gamma(\frac{3}{2}) = (\frac{1}{t_o\Gamma})^{1/2}\,\frac{\sqrt{\pi}}{2}$.
Then

$$\frac{1}{\Gamma t_o} \lesssim 1.3\left(\frac{80\,\text{eV}}{m_\nu}\right)^2 \qquad m_\nu \ll 1\,\text{MeV} \tag{7.68}$$

$$\frac{1}{\Gamma t_o} \lesssim 10^{-3}\frac{X_s}{\sqrt{c}}\left(\frac{m_\nu}{m_{Prot}}\right)^4 \qquad m_\nu \gg 1\,\text{MeV} \tag{7.69}$$

Further limits come from Nucleosynthesis[17]; results are summarized in Fig. 7.3.

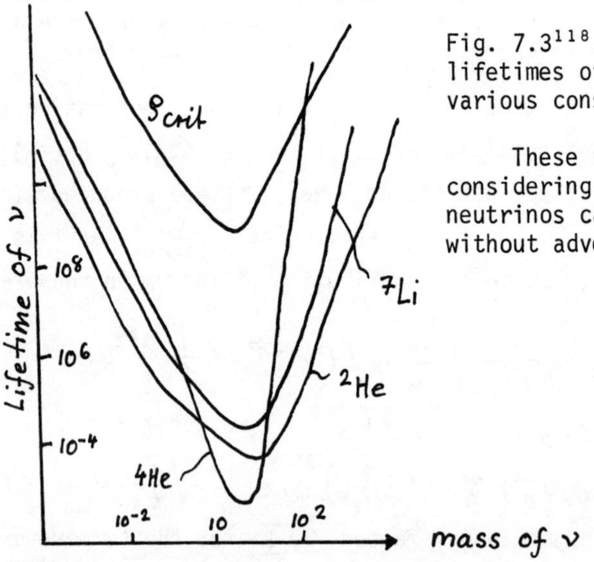

Fig. 7.3[118]. Upper limites on lifetimes of neutrinos from various considerations.

These results have led to considering models in which the neutrinos can decay fast, but without adverse effects[119].

ACKNOWLEDGEMENT

I thank the organizers of the School for their hospitality and all participants and other lecturers for creating a pleasant and interesting atmosphere.

I am indebted to K. Gabathuler, G. Zachek, V. Zachek and R.Frosch for patiently explaining to me the experimental aspects of neutrino masses and to G.Zachek for permitting using the illustrations in her thesis; and to N. Straumann and Ph. Jetzer for numerous discussions on the cosmological aspects; furthermore I thank B. Jost and A.Zehnder for discussions.

REFERENCES

1) See L.M. Brown, Physics Today, September 1978, p.23; W. Pauli, VII Congrès de Physique Solvay 1933 (Gauthier-Villars, Paris 1934), p. 324; the term "neutrino" is apparently due to Fermi, Ric. Sci., 4 (1933) 491; Nuov. Cim. II (1934) 1. A detailed history of weak interactions can be found in many textbooks; see also S.Focardi and R.A. Ricci, Riv. del Nuov. Cim. II (1983) 1.
2) Particle data booklet 1984; Rev. Mod. Phys. 56 (1984). Detailed references will be given later.
3) See B. Ovrut, these proceedings for an exposition of supersymmetry.
4) See W. Marciano, these proceedings for an exposition of gauge theories and of the standard models.
5) For a description of such theories, see M.A.B. Bég, these proceedings.
6) See H. Terezawa, these proceedings, for a description of such theories.
7) An exception is the model of L. Abbot and E. Farhi, Phys. Lett. 101B (1981) 69.
8) The transformation properties of spinors are treated in many textbooks. See e.g. J. Bjorken and S. Drell; Relativistic Quantum Mechanics + Field theory, Mc Graw Hill, New York 1964.
9) E. Majorana, Nuov. Cim.14 (1937) 171; G. Racah, Nuov. Cim. 14 (1937) 332. A Majorana spinor is neutral in the sense of e.g. the π^0.
10) For a summary of properties of Majorana spinors, see e.g. S.P. Rosen, Los Alamos preprint LA-UR-84-530.
11) The propagator for Majorana fields has been derived by many authors. See e.g. M. Doi et al. Progi. Theor. Phys. 66 (1981) 1739.
12) L. Wolfenstein, Phys. Lett. B 107 (1981) 77. See also: S.P. Rosen, Phys. Rev. D 29 (1984) 2535; D 30 (1984) 1995 (E), J. Bernabéu and P. Pascual, Barcelona Preprint; B. Kayser, Phys. Rev. D 30 (1984) 1023.
13) L. Wolfenstein, Nucl. Phys. B 186 (1981) 147; C.N. Leung and S.T. Petcor, Phys. Lett B 125 (1983) 461, S.M. Bilenky and B. Pontecorvo, Phys. Lett. B 102 (1981) 32; S. Petcor, Phys. Lett B 143 (1984) 175. The idea that ν_e,ν_μ could be combined into a four-spinor is old; see e.g. Y.S. Zeldovich, DAN SSS R 86 (1952) 505; E.J. Konopinsky and H. Mahmoud, Phys. Rev. 92 (1953) 1045; I Kawakami, Prog. theor. Phys. 19 (1958) 459; see also M. Nakagawa et al. Prog. theor. Phys. 30 (1963) 727.
14) L. Wolfenstein, ref. 13; see K. Fujikawa and R. Shrock, Phys. Rev. Lett. 45 (1980) 963.
15) D. Wyler and L. Wolfenstein, Nucl. Phys. B 218 (1983) 205.
16) More complete results can be found in B. Kayser, Phys. Rev. D 26 (1982) 1662; Phys. Rev. D 28 (1983) 2341.
17) S. Weinberg, Phys. Rev. Lett. 43 (1979) 1566.
18) H.M. Georgi and S.L. Glashow, Phys. Rev. Lett. 32 (1974) 438.
19) See, for example, R. Barbieri, D.V. Nanopoulos and D. Wyler, Phys. Lett. B 103 (1981) 433.

20) H.M. Georgi, Nucl. Phys. B 156 (1979) 126.

21) The question of neutrino masses is addressed in: J. Pati and A. Salam: Phys. Rev. Lett. 31 (1973) 661. For a review see R. Mohapatra, Maryland Preprint MD-DP-PP-840012. R.N.Mohapatra and G. Senjanovic, Phys. Rev. D 23 (1981) 165.

22) H. Georgi, in Particles and Fields (AIP Press, New York (1975)); H. Fritzsch and P. Minkowski, Ann. of Phys. 93 (1975) 193; H. Georgi and D.V. Nanopoulos, Nucl. Phys. B 155 (1976) 52.

23) For neutrino masses in SO(10) see: T. Yanagida, Proc. Workshop on Unified Theory and Baryon number in the Universe, Eds. O. Saweda and A. Sugamoto, KEK (1973); M. Gell-Mann, P. Ramond and R. Stansky, in Supergravity; Eds. P. van Nieuwenhuizen and D. Freedman (North-Holland, 1979) p. 317. See also P. Minkowski, Phys. Lett. 67B (1977) 421.

24) For a review of grand unified models. See e.g. P. Langacker, Phys. Rep. C 72 (1981) 185.

25) E. Witten, Phys. Lett. 91B (1980) 81.

26) M. Gronau, C.N. Leung and J.L. Rosner, Phys. Rev. D 29 (1984) 2539.

27) D. Wyler and L. Wolfenstein, ref. 15.

28) M. Roncadelli and D. Wyler, Phys. Lett. 133B (1983) 325. P. Roy and O. Shankar, Phys. Lett. 52 (1984) 713; 52 (1984) 2190 (E). The motivation for such models is the experimental indication for a neutrino mass (ref. 66) but absence of a Majorana mass (ref. 90).

29) R. Peccei and H. Quinn, Phys. Lett. 38 (1977) 1440; Phys. Rev. D 16 (1977) 1791.

30) P. Sikkivie, Phys. Rev. Lett. 49 (1982) 1549; A general discussion of cosmology of Goldstone bosons can be found in F. Wilczek, Erice Lecture, 1983, Santa Barbara preprint NSF-ITP-84-14. (1984).

31) Neutrino oscillations were first considered by B. Pontecorvo, Zh. Eksp. Teor., Fiz 33 (1957) 549 (JETP 6 (1957) 429;) 34 (1958), 247 (JETP 7 (1958) 172). Further early work: Z. Maki, M. Nakagawa and S. Sakata, Progr. theor. Phys. 28 (1962) 870, B. Pontecorvo, ibid. 53 (1967) 1717 (JETP 26 (1968) 984), V. Gribov and B. Pontecorvo, Phys. Lett. 28B, (1969) 495. A Review treating particularly this aspect of neutrino oscillations is S.M. Bilenki and B. Pontecorvo, Phys. Rep. 41C (1978) 225. More recent reviews are: H. Primakoff and S.P. Rosen, Ann. Rev. Nucl. Sci. 46 (1983) 1028; Ch. Baltay, Proc. Neutrino 81 Int. Conf. on Neutrino Physics and Astrophysics (ν 81) Maui, Hawaii; Ed. R.J. Cence, E. Ma and A. Roberts, Univ. Hawaii Press, I, (1981) 1; (experimental); P. Frampton and P. Vogel, Phys. Rep. 82 (1982) 33, M. Shaewitz, Proc. of the Cornell Lepton-Photon Conference 1983; B.A. Ljubimov, Proc. XXII Intntl. Conference on high Energy Physics, Leipzig, 1984 (experimental). A quantum mechanical discussion is given by B. Kayser, Phys. Rev. D 24 (1981) 110.

32) I thank M. Gronau for pointing out a possible difficulty with flavor conservation for Majorana particles. The general result is by S.L. Glashow and S. Weinberg, Phys. Rev. D 15 (1977) 1958.

33) Oscillations with both Majorana and Dirac Masses have been considered by: S.M. Bilenky, J. Hosek and S.T. Petcov, Phys. Lett. 94B (1980) 495; S.M. Bilenky and B. Pontecorvo, Phys. Lett. 102 (1981) 32; J. Maalampi and M. Roos, Phys. Lett. 146B (1984) 333; V. Barger, P. Langacker, J.P. Leveille and S. Pakvasa, Phys. Rev. Lett. 45 (1980) 692; S.M. Bilenky and B. Pontecorvo, Lett. Nuov. Cim. 17 (1976) 569.
34) M. Kobayashi and T. Maskawa, Progr. theor. Physics 49 (1973) 652.
35) See S.M. Bilenky et al., ref. 33.
36) T. Piran, Phys. Lett. 102B (1981) 669. N. Cabibbo, in Astrophys. and Elementary Particles, Common problems (Acad. Nat. d. Lincei, Roma 1980).
37) P. Frampton and P. Vogel, Op. cit.
38) F. Reines, H.W. Sobel and E. Pasierb. Phys. Rev. Lett. 45 (1980) 1307.
39) J.L. Vuilleumier et al., Phys. Lett. B114 (1982) 298.
40) See ref. 37 for a discussion of the problems involved as well as a list of references; see e.g. P. Vogel et al., Phys. Rev. C24 (1981) 1543; H.V. Klapdor and J. Metzinger, Phys. Rev. Lett. 48 (1981) 127.
41) G. Zachek, Ph. D. Thesis, 1984, Technical University of Munich.
42) K. Gabathuler et al., Phys. Lett. B 138 (1984) 449.
43) J.F. Cavaignac et al., Phys. Lett. B 148 (1984) 387.
44) N. Baker et al., Phys. Rev. Lett. 47 (1981) 1576.
45) N. Ushida et al., Phys. Rev. Lett. 47 (1981) 1694.
46) F. Dydak et al., Phys. Lett. B 134 (1984) 281. This is a ν_μ disappearance experiments with two distances.
47) I.E. Stockdale, Phys. Rev. Lett. 52 (1984) 1384.
48) F. Bergsma et al., Phys. Lett. B 142 (1984) 104; Earlier results on $\nu_\mu \to \nu_x$ include: J. Blietschau et al., Nucl. Phys. B 133 (1978) 205; S.E. Willis et al., Phys. Rev. Lett. 44 (1980) 522; P.N. Nemethy et al., Phys. Rev. D 23 (1981) 262.
49) M.Jonker et al., Phys. Lett. B 96 (1980) 435; P. Fritze et al.; ibid, 427; H. Abramovicz et al., Z. Phys. C (1981) 179.
50) J.N. Bahcall et al., Rev. Mod. Phys. 54 (1982) 767, B. Filippone and D. Schramm, Astrophys. J. 253 (1982) 393 give 7 $\pm$ 3 and 111 $\pm$ 13 for the total rates of Cl and Ga, respectively.
51) R. Davis, Jr., D.S. Harmer and K.C. Hoffman, Phys. Rev. Lett. 20 (1968) 1205; The newest result is in B.T. Cleveland, R. Davis and J.R. Rowley, in "Weak Interactions as probe of Unification", ed. by G. Collins, L. Chang and J. Fience (AIP conf. Proc. 72) p. 322.
52) This corresponds to 1.5 standard deviations from 1.
53) As discussed, Grand Unified theories might lead to very small neutrino masses. An application to the solar problem is in R. Barbieri, J.Ellis and M.K. Gaillard, Phys. Lett. B 90 (1980) 249.
54) For a recent calculation, see G. David, G. Marx, I. Ruff; Proc. (ν 84), Dortmund: Eötvös University preprint.
55) V.A. Kuzmin, Zh. Eksp. Theor. Fiz. 49 (1965) 1532.
56) M.A. Markov and I.M. Zheleznych, Nucl. Phys. 27 (1911) 385. See S.M. Bilenky and B. Pontecorvo, Op. cit.

57) S. Petcov, Yad. Fiz. 25 (1977) 641. Many authors have calculated (4.38) in models with heavy leptons (see S.M. Bilenky and B. Pontecorvo for references). A recent paper on the subject is T.P. Cheng and L.F. Li, Phys. Rev. Lett. 45 (1980) 1407.
58) L. Wolfenstein, Phys. Rev. D 17 (1978) 2369.
59) R. Abela et al., Phys. Lett. 146B (1984) 431. A previous bound used $K \to \Pi \mu \nu_\mu$ (A.R. Clark et al., Phys. Rev. D 9 (1974) 533).
60) R. Shrock, Phys. Lett. 96B (1980) 159; Phys. Rev. D 24 (1981) 1232, 1275.
61) R.C. Minehart et al., Phys. Rev. Lett. 52 (1984) 804.
62) R. Abela et al., Phys. Lett. B 105 (1981) 263.
63) See, e.g. J.M. Blatt and V.F. Weisskopf, Theoretical Nucl. Physics, Wiley and Sons, New York, 1952.
64) A nice discussion of the effects of $m_\nu \neq 0$ and $\Delta E \neq 0$ is in K.E. Bergkvist, Nucl. Phys. B 39 (1972) 317, 371.
65) G.C. Hanna and B. Pontecorvo, Phys. Rev. 75 (1949) 983; L.M. Langer and R.J.D. Moflat, Phys. Rev. 88, (1952) 689.
66) E.T. Tretjakov et al., Izv. Akad. Nauk SSSR, Ser. Fiz. 40 (1976) 20; V.A. Lubimov et al., Phys. Lett. 94B (1980) 266. The newest results are in V.A. Lubimov; op. cit. (ref. 31). A calculation of the valine effects has been done by Kaplan (above ref.).
67) A. de Rujula, Nucl. Phys. B 188 (1981) 414.
68) B. Jonson et al., Nucl. Phys. A 396 (1983) 449c; G. Beyer et al., Nucl. Phys. A 408 (1984).
69) J. Missimer, F. Scheck and R. Tegen, Nucl. Phys. B 188 (1981) 29.
70) Y. Asano et al., Phys. Lett. B 104 (1981) 84.
71) A.R. Clark, ref. 59.
72) D.A. Bryman et al., Phys. Rev. Lett. 50 (1983) 1546 (see also J. Ng. Phys. Lett. B 99 (1981) 53.
73) K. Heintze et al., Nucl. Phys. B 149 (1979) 365.
74) M. Göppert-Mayer, Phys. Rev. 48 (1935) 512. The $\Delta Z = -2$ is mentioned first in G. Dell'Antonio and E. Fiorini, Suppl. Nuov. Cim. 17 (1960) 132. For a recent review of results, see D.O. Caldwell, talk at the 1984 Leipzig Conference; Univ. of Calif. (Santa Barbara). Reviews with many references are P.G. Bizetti, Rev. del Nuov. Cim. 6 (1983) 1; H. Primakov and S.P.Rosen, Ann. Rev. Nucl. Sci. 31 (1981) 145, YuG. Zdesenko, Sov. J. Part. Nucl. II (1980) 542; D. Bryman and C. Piccioto.
75) B. Pontecorvo, Phys. Lett. 26B (1968) 630.
76) J. Abad, J.G. Esteve and A.F. Pacheco, Phys. Rev. D 30 (1984) 1488, J.N. Ng. and A.N. Kamal, Phys. Rev. D 18 (1978) 3412; A. Halprin, P. Minkowski, H. Primakoff and S.P. Rosen, Phys. Rev. D 13 (1976) 2567.
77) see ref. 12).
78) J.F. Nieves, Phys. Lett. B 147 (1984) 375; E. Takasugi, Phys. Lett. B 149 (1984) 375
79) R.N. Mohapatra and J.D. Vergados: Phys. Rev. Lett. 47 (1981) 1713.

80) L. Wolfenstein, Phys. Rev. D 26 (1982) 2507; J. Schechter and J.W.F. Valle, Phys. Rev. D 25 (1982) 2951; T.G. Rizzo and G. Senjanovic, Phys. Rev. D 25 (1982) 235.
81) J. Vergados: Telemark conference 1982 (ref. 68), C.E. Piccioto and M.S. Zahir.
82) J. Vergados, Phys. Rev. D 25 (1982).
83) M. Doi et al., Progr. Th. Phys. 66 (1981) 1739, 1765; 68 (1982) 347, 348 (E); 69 (1983) 602; 70 (1983) 1331, 1353.
84) C. Piccioto, Can. J. Phys. 55 (1978) 399.
85) T. Kirsten, H. Richter and E. Fessbender, Phys. Rev. Lett. 50 (1983) 474; Z. Phys. C 16 (1983) 189. The first paper lists the relevant references.
86) W.C. Haxton and G.J. Stephenson Jr., Pr. 250 (1982) 2360; Progr. in Part. and Nucl. Phys. 12, Pergamon Press 1984; M.K. Moe and D.D. Löwenthal, Phys. Rev. C 22 (1980) 2186.
87) P. Minkowski, Nucl. Phys. B 201 (1982) 269.
88) E. Hennecke et al., Phys. Rev. C 17 (1978) 1168.
89) E. Fiorini et al., Nuov. Cim. 13A (1973) 747; E. Bellotti et al., Phys. Lett. B 121 (1983) 72.
90) E. Bellotti et al., Phys. Lett. B 146 (1984) 450.
91) D. Caldwell et al., Leipzig Conference 1984.
92) J.J. Simpson et al., Phys. Rev. Lett. 53 (1984) 141.
93) H. Ejiri et al., Osaka Univ. Preprint.
94) A. Forster et al., Phys. Lett. B 138 (1984) 146.
95) F.T. Avignone et al., Phys. Rev. Lett. 50 (1983) 721.
96) Some Reviews which stress particularly this aspect are:
R. Cowsik, Moriond Proceedings 1981, Ed. Tran
J. Primack and G. Blumenthal, Moriond Proceedings 1983, Ed. Tran
N. Straumann, GIFT Seminar on Electroweak interactions (Spain) 1980
Some early work is: M. Markov, Phys. Rev. Lett. 10 (1984) 122
S. Gershtein and Y. Zeldovich, JETP Lett. 4 (1966) 407
N. Cabibbo, J. Bahcall and A. Yahil, Phys. Rev. Lett. 27 (1971) 757
M. Markov, Neutrino 174, (AIP Conf. Proc. New York 1974)
A. Dolgov and Y. Zeldovich, Rev. Mod. Phys. 53 (1981) 1
P. Hut and S. White, Nature 310 (1984) 637
97) S. Weinberg, Gravitation and Cosmology, Wiley, New York 1972.
98) A. Sandage and G. Tamman, Astrophys. J. 356 (1982) 339.
99) G. de Vaucouleurs, Nature 266 (1977) 126.
100) A. Sandage Astrophys. J. 173 (1972) 485.
101) R. Cowsik and J. McClelland, Phys. Rev. Lett. 29 (1972) 669; For an early paper, see A. Gerstein and Y. Zeldovich, Zh.Eks.Theor. Phiz. Pis'ma Red 4 (1966) 174.
102) G. Marx and A. Szalay, Proc. Neutrino 72 Budapest 1972.
103) B.W. Lee and S. Weinberg, Phys. Rev. Lett. 39 (1977) 165.
104) This treatment follows N. Straumann, Ref. 96;
105) V. Shvartsman, JETP Lett. 9 (1969) 184; G. Steigman, D. Schramm and J. Gunn, Phys. Lett. B 66 (1977) 202.
106) A. Rothenberg, Proc. Moriond Conf. 1984. This result comes from the recent Z, W runs at the SPS at CERN.

107) V. Rubin, Scientific American 248 (June 1983), p. 87).
108) R. Cowsik and J. McClelland, Astrophys. J. 180 (1973);
G. Marx and A. Szalay, Astron. and Astrophys. 49 (1976) 437.
109) J. Bond, G. Efstathiou and J. Silk, Phys. Rev. Lett. 45 (1980).
110) S. Tremaine and J. Gunn, Phys. Rev. Lett. 42 (1979) 407.
111) L. Landau and E. Lifshitz, Statistical Physics (Addison Wesely, 1969).
112) P. Shapiro et al., Astrophys. J. 275 (1983) 413;
J. Bond and A. Szalay, Astrophys. J. 274 (1983) 443.
113) see Cowsik, ref. a)
114) S. Falk and D. Schramm, Phys. Lett. 79B (1978) 511;
R. Cowsik, Phys. Rev. D 19 (1979) 2219.
115) A. De Rujula and S. Glashow, Phys. Rev. Lett. 45 (1980) 942.
116) D.A. Dicus, E. Kolb and V. Teplitz, Phys. Rev. Lett. 39 (1977) 168; Astrophys. J. 221 (1978) 327.
117) D.A. Dicus, E. Kolb, V. Teplitz and R. Wagoner, Phys. Rev. D 17, (1978) 1529.
118) S. Miyama and K. Sato, Prog. Th. Ohys. 60 (1978) 703.
119) J. Valle, Phys. Lett. B 131 (1983) 87; G. Gelmini and J. Valle, Phys. Lett. B 142 (1984) 181; A. Kumar and R. Mohapatra, Maryland preprint 85-44 (1984); G. Gelmini, D. Schramm and J. Valle, CERN preprint TH 3865 (1984).

LIST OF PARTICIPANTS

Manuel Avila
CIEA—IPN

Fernando A. Barrios
Instituto de Física—U.N.A.M.

Nora E. Bretón
CIEA—IPN

Eduardo Cantoral
Universidad Autónoma de Puebla

Fernando Chaos
E.S.F.M.—IPN

Pablo Chauvet
U. Autónoma Metropolitana
Iztapalapa

Efraín R. Chávez
Instituto de Física—U.N.A.M.

Arturo Cisneros
E.S.F.M.—IPN

Juan Carlos D'Olivo
Centro de Estudios Nucleares
U.N.A.M.

Joaquín Escalona
Instituto de Física—U.N.A.M.

David José Fernández
CIEA–IPN

Augusto García
E.S.F.M.—IPN

Ignacio Luis Garzón
Instituto de Física—U.N.A.M.

Catalina Guerra
Instituto de Física—U.N.A.M.

Albino Hernández
CIEA—IPN

Rodrigo Huerta
CIEA—IPN

Sara Rebeca Juárez
E.S.F.M.—IPN

Jaime Keller
Facultad de Estudios Superiores
Cuautitlán—U.N.A.M.

José Luis López
E.S.F.M.—IPN

Gabriel López
CIEA—IPN

Gerardo Loyola
Instituto de Física—U.N.A.M.

José Luis Lucio
CIEA—IPN

Ma. Alejandrina Martínez
CIEA—IPN

Alfonso Martínez
E.S.F.M.—IPN

Mario Alberto Maya
Universidad Autónoma de Puebla

Héctor V. Méndez
CIEA—IPN

Gerardo Moreno
CIEA—Ipn

Matías Moreno
Instituto de Física—U.N.A.M.

Ekkehard Nowotny
U. Autónoma Metropolitana
Iztapalapa

Alfredo Peña
Facultad de Estudios Superiores
Cuautitlán—U.N.A.M.

Miguel A. Pérez
CIEA—IPN

Raúl Pérez
Universidad Autónoma de Puebla

Octavio Pimentel
U. Autónoma Metropolitana
Iztapalapa

Alfonso Quejeiro
E.S.F.M.—IPN

Alejandro Ramírez
Instituto de Física—U.N.A.M.

Manuel Rivera
E.S.F.M.—IPN

Ulises Rodríguez
Instituto de Física—U.N.A.M.

Sergio Saavedra
Instituto de Física—U.N.A.M.

Patricia Salas
Instituto de Física—U.N.A.M.

Humberto A. Salazar
Universidad Autónoma de Puebla

Miguel Sánchez
Instituto de Física—U.N.A.M.

Miguel Socolovsky
CIEA—IPN

Miguel Angel Soriano
Universidad Autónoma de Puebla

Juan Suro
Instituto de Física—U.N.A.M.

Luis Urrutia
Centro de Estudios Nucleares
U.N.A.M.

Carlos Villarreal
Instituto de Física—U.N.A.M.

Luis Manuel Villaseñor
CIEA—IPN

Arturo Zentella
Instituto de Física—U.N.A.M.

Arnulfo Zepeda
CIEA—IPN

LIST OF SEMINARS

On the meaning of blacker than black .... E. Cantoral

Hard photon bremsstrahlung corrections to semileptonic baryon decays .... R. Juarez

Manifestation of the SUSY content of the proton wave function in $p\bar{p}$ collisions .... J. L. Lucio

Triviality of the $\lambda\phi^4$ theory and the t quark mass .... E. Masso

Nonlinear realization of supersymmetry .... J. Millan

Some applications of the path integral in quantum mechanics .... L. Plis

The scalar field and the mass acquisition in the Weinberg–Salam theory .... D. Rapaport

Manifestations of supersymmetry in nuclear physics .... L. Urrutia

AIP Conference Proceedings

		L.C. Number	ISBN
No. 1	Feedback and Dynamic Control of Plasmas – 1970	70-141596	0-88318-100-2
No. 2	Particles and Fields – 1971 (Rochester)	71-184662	0-88318-101-0
No. 3	Thermal Expansion – 1971 (Corning)	72-76970	0-88318-102-9
No. 4	Superconductivity in *d*- and *f*-Band Metals (Rochester, 1971)	74-18879	0-88318-103-7
No. 5	Magnetism and Magnetic Materials – 1971 (2 parts) (Chicago)	59-2468	0-88318-104-5
No. 6	Particle Physics (Irvine, 1971)	72-81239	0-88318-105-3
No. 7	Exploring the History of Nuclear Physics – 1972	72-81883	0-88318-106-1
No. 8	Experimental Meson Spectroscopy –1972	72-88226	0-88318-107-X
No. 9	Cyclotrons – 1972 (Vancouver)	72-92798	0-88318-108-8
No. 10	Magnetism and Magnetic Materials – 1972	72-623469	0-88318-109-6
No. 11	Transport Phenomena – 1973 (Brown University Conference)	73-80682	0-88318-110-X
No. 12	Experiments on High Energy Particle Collisions – 1973 (Vanderbilt Conference)	73-81705	0-88318-111–8
No. 13	π-π Scattering – 1973 (Tallahassee Conference)	73-81704	0-88318-112-6
No. 14	Particles and Fields – 1973 (APS/DPF Berkeley)	73-91923	0-88318-113-4
No. 15	High Energy Collisions – 1973 (Stony Brook)	73-92324	0-88318-114-2
No. 16	Causality and Physical Theories (Wayne State University, 1973)	73-93420	0-88318-115-0
No. 17	Thermal Expansion – 1973 (Lake of the Ozarks)	73-94415	0-88318-116-9
No. 18	Magnetism and Magnetic Materials – 1973 (2 parts) (Boston)	59-2468	0-88318-117-7
No. 19	Physics and the Energy Problem – 1974 (APS Chicago)	73-94416	0-88318-118-5
No. 20	Tetrahedrally Bonded Amorphous Semiconductors (Yorktown Heights, 1974)	74-80145	0-88318-119-3
No. 21	Experimental Meson Spectroscopy – 1974 (Boston)	74-82628	0-88318-120-7
No. 22	Neutrinos – 1974 (Philadelphia)	74-82413	0-88318-121-5
No. 23	Particles and Fields – 1974 (APS/DPF Williamsburg)	74-27575	0-88318-122-3
No. 24	Magnetism and Magnetic Materials – 1974 (20th Annual Conference, San Francisco)	75-2647	0-88318-123-1

No. 25	Efficient Use of Energy (The APS Studies on the Technical Aspects of the More Efficient Use of Energy)	75-18227	0-88318-124-X
No. 26	High-Energy Physics and Nuclear Structure – 1975 (Santa Fe and Los Alamos)	75-26411	0-88318-125-8
No. 27	Topics in Statistical Mechanics and Biophysics: A Memorial to Julius L. Jackson (Wayne State University, 1975)	75-36309	0-88318-126-6
No. 28	Physics and Our World: A Symposium in Honor of Victor F. Weisskopf (M.I.T., 1974)	76-7207	0-88318-127-4
No. 29	Magnetism and Magnetic Materials – 1975 (21st Annual Conference, Philadelphia)	76-10931	0-88318-128-2
No. 30	Particle Searches and Discoveries – 1976 (Vanderbilt Conference)	76-19949	0-88318-129-0
No. 31	Structure and Excitations of Amorphous Solids (Williamsburg, VA, 1976)	76-22279	0-88318-130-4
No. 32	Materials Technology – 1976 (APS New York Meeting)	76-27967	0-88318-131-2
No. 33	Meson-Nuclear Physics – 1976 (Carnegie-Mellon Conference)	76-26811	0-88318-132-0
No. 34	Magnetism and Magnetic Materials – 1976 (Joint MMM-Intermag Conference, Pittsburgh)	76-47106	0-88318-133-9
No. 35	High Energy Physics with Polarized Beams and Targets (Argonne, 1976)	76-50181	0-88318-134-7
No. 36	Momentum Wave Functions – 1976 (Indiana University)	77-82145	0-88318-135-5
No. 37	Weak Interaction Physics – 1977 (Indiana University)	77-83344	0-88318-136-3
No. 38	Workshop on New Directions in Mossbauer Spectroscopy (Argonne, 1977)	77-90635	0-88318-137-1
No. 39	Physics Careers, Employment and Education (Penn State, 1977)	77-94053	0-88318-138-X
No. 40	Electrical Transport and Optical Properties of Inhomogeneous Media (Ohio State University, 1977)	78-54319	0-88318-139-8
No. 41	Nucleon-Nucleon Interactions – 1977 (Vancouver)	78-54249	0-88318-140-1
No. 42	Higher Energy Polarized Proton Beams (Ann Arbor, 1977)	78-55682	0-88318-141-X
No. 43	Particles and Fields – 1977 (APS/DPF, Argonne)	78-55683	0-88318-142-8
No. 44	Future Trends in Superconductive Electronics (Charlottesville, 1978)	77-9240	0-88318-143-6
No. 45	New Results in High Energy Physics – 1978 (Vanderbilt Conference)	78-67196	0-88318-144-4

No. 46	Topics in Nonlinear Dynamics (La Jolla Institute)	78-57870	0-88318-145-2
No. 47	Clustering Aspects of Nuclear Structure and Nuclear Reactions (Winnepeg, 1978)	78-64942	0-88318-146-0
No. 48	Current Trends in the Theory of Fields (Tallahassee, 1978)	78-72948	0-88318-147-9
No. 49	Cosmic Rays and Particle Physics – 1978 (Bartol Conference)	79-50489	0-88318-148-7
No. 50	Laser-Solid Interactions and Laser Processing – 1978 (Boston)	79-51564	0-88318-149-5
No. 51	High Energy Physics with Polarized Beams and Polarized Targets (Argonne, 1978)	79-64565	0-88318-150-9
No. 52	Long-Distance Neutrino Detection – 1978 (C.L. Cowan Memorial Symposium)	79-52078	0-88318-151-7
No. 53	Modulated Structures – 1979 (Kailua Kona, Hawaii)	79-53846	0-88318-152-5
No. 54	Meson-Nuclear Physics – 1979 (Houston)	79-53978	0-88318-153-3
No. 55	Quantum Chromodynamics (La Jolla, 1978)	79-54969	0-88318-154-1
No. 56	Particle Acceleration Mechanisms in Astrophysics (La Jolla, 1979)	79-55844	0-88318-155-X
No. 57	Nonlinear Dynamics and the Beam-Beam Interaction (Brookhaven, 1979)	79-57341	0-88318-156-8
No. 58	Inhomogeneous Superconductors – 1979 (Berkeley Springs, W.V.)	79-57620	0-88318-157-6
No. 59	Particles and Fields – 1979 (APS/DPF Montreal)	80-66631	0-88318-158-4
No. 60	History of the ZGS (Argonne, 1979)	80-67694	0-88318-159-2
No. 61	Aspects of the Kinetics and Dynamics of Surface Reactions (La Jolla Institute, 1979)	80-68004	0-88318-160-6
No. 62	High Energy e^+e^- Interactions (Vanderbilt, 1980)	80-53377	0-88318-161-4
No. 63	Supernovae Spectra (La Jolla, 1980)	80-70019	0-88318-162-2
No. 64	Laboratory EXAFS Facilities – 1980 (Univ. of Washington)	80-70579	0-88318-163-0
No. 65	Optics in Four Dimensions – 1980 (ICO, Ensenada)	80-70771	0-88318-164-9
No. 66	Physics in the Automotive Industry – 1980 (APS/AAPT Topical Conference)	80-70987	0-88318-165-7
No. 67	Experimental Meson Spectroscopy – 1980 (Sixth International Conference, Brookhaven)	80-71123	0-88318-166-5
No. 68	High Energy Physics – 1980 (XX International Conference, Madison)	81-65032	0-88318-167-3
No. 69	Polarization Phenomena in Nuclear Physics – 1980 (Fifth International Symposium, Santa Fe)	81-65107	0-88318-168-1

No. 70	Chemistry and Physics of Coal Utilization – 1980 (APS, Morgantown)	81-65106	0-88318-169-X
No. 71	Group Theory and its Applications in Physics – 1980 (Latin American School of Physics, Mexico City)	81-66132	0-88318-170-3
No. 72	Weak Interactions as a Probe of Unification (Virginia Polytechnic Institute – 1980)	81-67184	0-88318-171-1
No. 73	Tetrahedrally Bonded Amorphous Semiconductors (Carefree, Arizona, 1981)	81-67419	0-88318-172-X
No. 74	Perturbative Quantum Chromodynamics (Tallahassee, 1981)	81-70372	0-88318-173-8
No. 75	Low Energy X-Ray Diagnostics – 1981 (Monterey)	81-69841	0-88318-174-6
No. 76	Nonlinear Properties of Internal Waves (La Jolla Institute, 1981)	81-71062	0-88318-175-4
No. 77	Gamma Ray Transients and Related Astrophysical Phenomena (La Jolla Institute, 1981)	81-71543	0-88318-176-2
No. 78	Shock Waves in Condensed Mater – 1981 (Menlo Park)	82-70014	0-88318-177-0
No. 79	Pion Production and Absorption in Nuclei – 1981 (Indiana University Cyclotron Facility)	82-70678	0-88318-178-9
No. 80	Polarized Proton Ion Sources (Ann Arbor, 1981)	82-71025	0-88318-179-7
No. 81	Particles and Fields –1981: Testing the Standard Model (APS/DPF, Santa Cruz)	82-71156	0-88318-180-0
No. 82	Interpretation of Climate and Photochemical Models, Ozone and Temperature Measurements (La Jolla Institute, 1981)	82-71345	0-88318-181-9
No. 83	The Galactic Center (Cal. Inst. of Tech., 1982)	82-71635	0-88318-182-7
No. 84	Physics in the Steel Industry (APS/AISI, Lehigh University, 1981)	82-72033	0-88318-183-5
No. 85	Proton-Antiproton Collider Physics –1981 (Madison, Wisconsin)	82-72141	0-88318-184-3
No. 86	Momentum Wave Functions – 1982 (Adelaide, Australia)	82-72375	0-88318-185-1
No. 87	Physics of High Energy Particle Accelerators (Fermilab Summer School, 1981)	82-72421	0-88318-186-X
No. 88	Mathematical Methods in Hydrodynamics and Integrability in Dynamical Systems (La Jolla Institute, 1981)	82-72462	0-88318-187-8
No. 89	Neutron Scattering – 1981 (Argonne National Laboratory)	82-73094	0-88318-188-6
No. 90	Laser Techniques for Extreme Ultraviolt Spectroscopy (Boulder, 1982)	82-73205	0-88318-189-4

No. 91	Laser Acceleration of Particles (Los Alamos, 1982)	82-73361	0-88318-190-8
No. 92	The State of Particle Accelerators and High Energy Physics (Fermilab, 1981)	82-73861	0-88318-191-6
No. 93	Novel Results in Particle Physics (Vanderbilt, 1982)	82-73954	0-88318-192-4
No. 94	X-Ray and Atomic Inner-Shell Physics – 1982 (International Conference, U. of Oregon)	82-74075	0-88318-193-2
No. 95	High Energy Spin Physics – 1982 (Brookhaven National Laboratory)	83-70154	0-88318-194-0
No. 96	Science Underground (Los Alamos, 1982)	83-70377	0-88318-195-9
No. 97	The Interaction Between Medium Energy Nucleons in Nuclei – 1982 (Indiana University)	83-70649	0-88318-196-7
No. 98	Particles and Fields – 1982 (APS/DPF University of Maryland)	83-70807	0-88318-197-5
No. 99	Neutrino Mass and Gauge Structure of Weak Interactions (Telemark, 1982)	83-71072	0-88318-198-3
No. 100	Excimer Lasers – 1983 (OSA, Lake Tahoe, Nevada)	83-71437	0-88318-199-1
No. 101	Positron-Electron Pairs in Astrophysics (Goddard Space Flight Center, 1983)	83-71926	0-88318-200-9
No. 102	Intense Medium Energy Sources of Strangeness (UC-Sant Cruz, 1983)	83-72261	0-88318-201-7
No. 103	Quantum Fluids and Solids – 1983 (Sanibel Island, Florida)	83-72440	0-88318-202-5
No. 104	Physics, Technology and the Nuclear Arms Race (APS Baltimore –1983)	83-72533	0-88318-203-3
No. 105	Physics of High Energy Particle Accelerators (SLAC Summer School, 1982)	83-72986	0-88318-304-8
No. 106	Predictability of Fluid Motions (La Jolla Institute, 1983)	83-73641	0-88318-305-6
No. 107	Physics and Chemistry of Porous Media (Schlumberger-Doll Research, 1983)	83-73640	0-88318-306-4
No. 108	The Time Projection Chamber (TRIUMF, Vancouver, 1983)	83-83445	0-88318-307-2
No. 109	Random Walks and Their Applications in the Physical and Biological Sciences (NBS/La Jolla Institute, 1982)	84-70208	0-88318-308-0
No. 110	Hadron Substructure in Nuclear Physics (Indiana University, 1983)	84-70165	0-88318-309-9
No. 111	Production and Neutralization of Negative Ions and Beams (3rd Int'l Symposium, Brookhaven, 1983)	84-70379	0-88318-310-2

No. 112	Particles and Fields – 1983 (APS/DPF, Blacksburg, VA)	84-70378	0-88318-311-0
No. 113	Experimental Meson Spectroscopy – 1983 (Seventh International Conference, Brookhaven)	84-70910	0-88318-312-9
No. 114	Low Energy Tests of Conservation Laws in Particle Physics (Blacksburg, VA, 1983)	84-71157	0-88318-313-7
No. 115	High Energy Transients in Astrophysics (Santa Cruz, CA, 1983)	84-71205	0-88318-314-5
No. 116	Problems in Unification and Supergravity (La Jolla Institute, 1983)	84-71246	0-88318-315-3
No. 117	Polarized Proton Ion Sources (TRIUMF, Vancouver, 1983)	84-71235	0-88318-316-1
No. 118	Free Electron Generation of Extreme Ultraviolet Coherent Radiation (Brookhaven/OSA, 1983)	84-71539	0-88318-317-X
No. 119	Laser Techniques in the Extreme Ultraviolet (OSA, Boulder, Colorado, 1984)	84-72128	0-88318-318-8
No. 120	Optical Effects in Amorphous Semiconductors (Snowbird, Utah, 1984)	84-72419	0-88318-319-6
No. 121	High Energy e^+e^- Interactions (Vanderbilt, 1984)	84-72632	0-88318-320-X
No. 122	The Physics of VLSI (Xerox, Palo Alto, 1984)	84-72729	0-88318-321-8
No. 123	Intersections Between Particle and Nuclear Physics (Steamboat Springs, 1984)	84-72790	0-88318-322-6
No. 124	Neutron-Nucleus Collisions – A Probe of Nuclear Structure (Burr Oak State Park - 1984)	84-73216	0-88318-323-4
No. 125	Capture Gamma-Ray Spectroscopy and Related Topics – 1984 (Internat. Symposium, Knoxville)	84-73303	0-88318-324-2
No. 126	Solar Neutrinos and Neutrino Astronomy (Homestake, 1984)	84-63143	0-88318-325-0
No. 127	Physics of High Energy Particle Accelerators (BNL/SUNY Summer School, 1983)	85-70057	0-88318-326-9
No. 128	Nuclear Physics with Stored, Cooled Beams (McCormick's Creek State Park, Indiana, 1984)	85-71167	0-88318-327-7
No. 129	Radiofrequency Plasma Heating (Sixth Topical Conference, Callaway Gardens, GA, 1985)	85-48027	0-88318-328-5
No. 130	Laser Acceleration of Particles (Malibu, California, 1985)	85-48028	0-88318-329-3
No. 131	Workshop on Polarized ^{3}He Beams and Targets (Princeton, New Jersey, 1984)	85-48026	0-88318-330-7
No. 132	Hadron Spectroscopy–1985 (International Conference, Univ. of Maryland)	85-72537	0-88318-331-5

No. 133	Hadronic Probes and Nuclear Interactions (Arizona State University, 1985)	85-72638	0-88318-332-3
No. 134	The State of High Energy Physics (BNL/SUNY Summer School, 1983)	85-73170	0-88318-333-1
No. 135	Energy Sources: Conservation and Renewables (APS, Washington, DC, 1985)	85-73019	0-88318-334-X
No. 136	Atomic Theory Workshop on Relativistic and QED Effects in Heavy Atoms	85-73790	0-88318-335-8
No. 137	Polymer-Flow Interaction (La Jolla Institute, 1985)	85-73915	0-88318-336-6
No. 138	Frontiers in Electronic Materials and Processing (Houston, TX, 1985)	86-70108	0-88318-337-4
No. 139	High-Current, High-Brightness, and High-Duty Factor Ion Injectors (La Jolla Institute, 1985)	86-70245	0-88318-338-2
No. 140	Boron-Rich Solids (Albuquerque, NM, 1985)	86-70246	0-88318-339-0
No. 141	Gamma-Ray Bursts (Stanford, CA, 1984)	86-70761	0-88318-340-4
No. 142	Nuclear Structure at High Spin, Excitation, and Momentum Transfer (Indiana University, 1985)	86-70837	0-88318-341-2